Fire and Fauna

Sponsored by Texas Research Institute
for Environmental Studies,
Sam Houston State University
William I. Lutterschmidt and
Brian R. Chapman, General Editors

To Anne,
Through a window, or on a trail: Enjoy Nature. and many thanks for your encouragement!

Fire and Fauna

Tales of a Life Untamed

JOAN E. BERISH

Joan E. Berish

Texas A&M University Press   College Station

Copyright © 2019
by Joan E. Berish
All rights reserved
First edition

This paper meets the requirements
of ANSI/NISO Z39.48-1992 (Permanence of Paper).
Binding materials have been chosen for durability.
Manufactured in the United States of America

Library of Congress Cataloging-in-Publication Data

Names: Berish, Joan E., 1950- author.
Title: Fire and fauna : tales of a life untamed / Joan E Berish.
Other titles: Integrative natural history series.
Description: First edition. | College Station : Texas A&M University Press,
　[2019] | Series: Integrative natural history series | Includes index.
Identifiers: LCCN 2019021682 | ISBN 9781623498313 (cloth) | ISBN
　9781623498320 (ebook)
Subjects: LCSH: Berish, Joan E., 1950- | Zoologists—Florida—Biography. |
　Wildlife conservationists—Florida—Biography. | Gopher
　tortoise—Conservation—Florida. | Wildlife conservation—Florida.
Classification: LCC QL31.B435 A3 2019 | DDC 590.92 [B]—dc23
LC record available at https://lccn.loc.gov/2019021682

*To my angel mother, Verna,
whose unwavering encouragement
gave me wings*

*To my late amigo, Ray,
who gave gopher tortoises a voice*

*To my husband, Dave,
who gave this wild biologist his love
and steadfast support*

Contents

A gallery of photographs follow page 130.

Prologue   ix
Acknowledgments   xi

PART 1. Evolution of a "Wild Biologist"   1

   Chapter 1. Child of the Fields and Forests   2
   Chapter 2. A Horse, a Horse: My Kingdom (Lunch Money)
      for a Horse   10
   Chapter 3. Would Dog Bite (Mail)Woman?   18
   Chapter 4. Going Nuclear at the Veterans Administration   25
   Chapter 5. Canines and Felines and . . . Willie Nelson?   31
   Chapter 6. Whirlybirds and Conflagrations: My Summer as a
      Helitack Firefighter   36

PART 2. Metamorphosis: From Snake Lady to Gopher Queen   57

   Chapter 7. Indigo Daze: The Snake Lady Goes Down to
      Georgia   59
   Chapter 8. Ancient Dunes, Black Holes, and Burrowing
      Turtles   95
   Chapter 9. A New Life Partner and Many More Moons of Gopher
      Wrangling   131
   Chapter 10. Rattlers and Softshells and Bears, Oh My!   167
   Chapter 11. Canines in Conservation and in My Heart   178

PART 3. Modifications to My Home Range, Both Temporary and
     Permanent   191

   Chapter 12. Misadventures on Desert Trails and High Seas   193
   Chapter 13. My Return to the Land of Enchantment   210

Index   231

Prologue

The Cessna 152 bucked and bounced in the late-afternoon thermals spawned by the staggering June heat. Although north-central Florida is rarely cool during early June, this 1985 heat wave was one for the record books. The air temperature was over 100°F with humidity so intense I felt as though I needed gills to breathe. Undeterred, my buddy and new pilot, Ab, and I duct-taped an antenna to the wing strut. The H-shaped antenna was designed to track wildlife with radio transmitters attached. When we had completed the job and climbed into the two-seater, we were hit with a cockpit temperature of 140°F. We were on a mission to find #29, a subadult gopher tortoise affectionately dubbed Dufus for its seemingly odd behavior and comic antics. Dufus was part of a tortoise home range study I was conducting as a research biologist for the Florida Game and Fresh Water Fish Commission.

Although I generally didn't name my study tortoises, Dufus was an exception. This youngster, whom I had first captured and marked in 1982, was proving to be one troublesome, but ultimately enlightening, burrowing turtle. When days of ground tracking in the searing heat failed to locate #29, I decided we needed to go higher, as in airplane high at five hundred feet, to try to get a radio signal. Now I was questioning that decision as the plane sustained gut-wrenching and somewhat terrifying aerial bumps. The terror factor increased tenfold when the plane suddenly stalled. Our regular, experienced pilot later noted that we never should have taken off in the first place, considering the extreme heat and the fact that my pilot was only recently licensed. Moreover, radio-tracking flights generally fly relatively low and slowly as the aircraft circles the target area to try to pick up beeps emitted by the transmitter on a specific radio frequency. Whatever the reason for the stall, it was undeniably an "Oh my God, I'm going to die" few moments that later haunted my dreams.

Luckily, the plane's engine kicked back into non-life-threatening flight mode, and I was able to focus on a faint signal coming through my earphones. The signal was not originating from my pine plantation study site but instead from a pasture quite a way to the southeast. Needless to say, our tenacious but naïve efforts could have gotten us killed, and I was therefore at least threefold thankful: to be alive; to have access to an airsickness bag (how fun is that?); and to have found the wayward, wee Dufus. This young tortoise had made a huge move. His journey into the unknown resonated with me, as I, too, had made moves that yielded unexpected adventures and memorable stories.

Acknowledgments

Writing this memoir required me to stay in my "burrow" for days at a time, but I was only a computer click away from considerable outside assistance to trim and hone the various drafts. First and foremost, I want to thank my amazing, animal-loving editor, Pam Sourelis. Not only are her editing skills outstanding, but she truly made me think about how other animal enthusiasts might perceive certain parts of the book. Laboratory research, veterinary medicine, and wildlife biology sometimes require undertaking difficult procedures with animals; I wanted to portray these honestly but to also emphasize the care and dedication involved. Pam helped me find that balance.

My earliest long draft was reviewed by ecologist Kathy Freas, and I greatly appreciate her suggestions and encouragement. An anonymous reviewer and my wildlife colleague Mary Anne Young provided solid critiques and gave me specific insights on how best to improve a later draft. Dog trainer Cynthia Angevine reviewed the canine chapter and helped me condense and sharpen the text. Nan Smith and Cliff Leonard graciously provided eleventh-hour photographic assistance.

Molly Jurhill has been my champion and cheerleader for nearly five decades of this wild journey, especially when I struggled to find jobs working with animals in New Mexico during the 1970s. And I owe a debt of gratitude to Deborah Burr for keeping the gopher tortoise conservation torch burning brightly in Florida.

A chance meeting at the 2017 Wildlife Society conference in Albuquerque with then acquisitions editor Stacy Eisenstark led me to send a query letter to Texas A&M University Press. I will always be grateful for Stacy's assistance.

Finally, I want to thank my husband, Dave, for both his patience and proofreading skills during this lengthy endeavor and literary journey. I'm sure he lost count of the times he heard: "I'm work ... ing!" He and all my friends near and far have been incredibly supportive as I trekked along this uphill, switchback trail toward publication.

Fire and Fauna

1

Evolution of a "Wild Biologist"

Keep thou thy dreams—the tissue of all wings is woven first of them.

—VIRNA SHEARD

1
Child of the Fields and Forests

Critters and a Creek

As I reflect on my life and career, I realize that I was continually on a quest. My desires were basically threefold: to be immersed in the world of animals and nature; to seek out adventures of many types, especially those in the great outdoors; and to satiate my innate curiosity and thirst for knowledge. The love of animals and nature guided me to a career working with wildlife; the need for adventure pushed me toward field biology; and my curious mind was a perfect fit for the job of researcher. As in anyone's life, there were side paths that did not contribute to this quest, but, amazingly, most of my early training and travails directly or indirectly led to my becoming a wildlife biologist.

I am told that one of the first words I uttered as a towheaded toddler was "pij." Not "Mommy" or "Daddy" or any number of other first words that are lovingly recorded by parents; instead, I apparently was intrigued with the pigeons that our neighbors were raising. My mother, Verna, a former home economics major, dietician, and then housewife of the 1950s, claimed that my innate love of all animals must have happened because I was a throwback to my great-uncle Pete in Norway. Apparently, Uncle Pete was a horseman and critter enthusiast. Alas, I never met Uncle Pete or any of my other Scandinavian ancestors in the old country, but my mom's explanation always seemed to make sense to me.

I have often wondered what combination of heredity and upbringing (the proverbial "nature vs. nurture" debate) initially prompted a little girl, especially one raised by such a traditionally feminine role model, to follow an early childhood path that was so wholeheartedly oriented toward animals and the outdoors.

What was the impetus of my love for animals and my steadfast determination to be among them? My father, Carroll ("Smitty"), taught my brothers and me to respect wild creatures. For example, we didn't kill snakes that slithered onto our acre of land in rural Maryland, and we enjoyed feeding the juncos and other wild birds. My two older siblings, Donnie and Davie, loved our dogs and cats and enjoyed fishing and playing outdoor war games and sports; however, neither brother ever really embraced the wild world later in life the way I did. Yet perhaps there was a genetic connection, because Donnie had an early interest in birds and collected their nests before he embarked on an army career. Today, he resides on a Texas ranchette and raises goats, rabbits, doves, ducks, and chickens, what I call Old Donald's farm. Davie went on to get a doctorate in neurobiology after working as a physicist. His PhD was on fly vision, and he wryly noted that dog feces provided him with a reliable source of these winged insects.

In any era, young children often express a love of both stuffed and real animals. Was I unusual for a girl child of the 1950s? I liked dolls and playing dress-up, but my passion was for wild places and creatures. I suppose I would have been considered a tomboy, growing up with brothers and exploring the fields, forests, and especially the creek that ran near our homestead. That creek held a powerful allure for me and was often the reason for my getting into trouble with my parents when I came home with wet clothes. Physical coordination has never been my strong suit, and I seemed to fall into the creek on a regular basis. In retrospect, I think the creek represented adventure: Where did it go once it left my neck of the woods? When I got a bit older, my brothers and their friends would sometimes let me tag along on their farther wanderings to Frog Island (presumably named because of the many amphibians there). I caught frogs and toads, which I kept for short periods; all were named Hoppy. And, prophetically, I also caught and kept box turtles, all of whom were named Dumpy.

Besides our beloved dogs (two consecutive Woofys and then little Randy), we had a quirky, gray-and-white mama cat named Mitzi. Reflecting my early admiration for the Wild West, I creatively named some of her kittens after 1950s cowboy heroes: Bret, Bart, Cheyenne, and Sugarfoot. On one occasion that we kids found hilarious, Mitzi covered up one of my mom's famous Swedish meatballs with dirt, as if it were a round piece of poop. My mother was not amused. I also briefly had a white rabbit named Thumper who lived in a cage on our land. I would take him out into the soft grass to lovingly play with him. But one day when I went to cuddle with Thumper, he uncharacteristically struggled and dug his sharp claws into my flesh. This aberrant behavior went on for days, much to my dismay (why didn't he love me anymore?)—until finally, my brothers told me that Dad had secretly replaced the rabbit because the original Thumper had been consumed by a fox. I suppose the secret was to spare me the loss of my pet. However, the replacement rabbit was eventually sold or given away because it never accepted my touch.

My early interest in animals was not confined to dogs, cats, rabbits, frogs, and turtles. The incipient scientist was emerging, as evidenced by my bringing home a fox skull, aptly named Stinky. Foxes, preferably alive with their impressive rusty or smoky coats and bushy tails, have always fascinated me. My mother, however, was not impressed with this discovery, and Stinky didn't occupy a space in our yard for long. But she was even less enamored with the pinkie rats that I toted home from some woodland haunt. I proudly took them to school the next day for show and tell (the bus driver also was not pleased) and then left them home for Mom to tend while I was in class. Surprisingly (to me), they all supposedly died shortly thereafter and were buried in the critter graveyard on our property. I will never know if they succumbed because they were too young to be weaned or if foul play was involved.

I also went through a goldfish stage, and not surprisingly, there was frequent mortality. My favorites were Comanche and Little Vic (named after horses), one a typical golden hue and the other a small red-and-white pinto fish. Because most of our finned critter casualties were unceremoniously flushed down the toilet, I felt that Comanche and Little Vic deserved better. So one day I snuck down to the creek and joyously released them to the wild. A child's perspective is very different from that of a professional wildlife biologist, and I, of course, eventually learned that this was not an ecologically responsible act. Exotic flora and fauna can be destructive and even devastating to wildlife and their habitats. That said, my guess is that these little fish did not survive long in their new home.

Although most insects weren't necessarily an early interest, I did maintain a butterfly collection of spectacular swallowtails and monarchs, and my friends and I spent many enjoyable hours catching fireflies in the gloaming of a summer's eve. Considerably less enjoyable were the pervasive ticks that haunted the unmowed, weed-dominated, edges of our acre. As a so-called tomboy, I played football, softball, soccer, and intense badminton with my brothers, one of whom had a short temper. One day when my parents were out of town and my neighbor/godmother was in charge of me, she got the unpleasant task of extracting eighteen ticks from my head after Davie shoved me into the weeds during a touch football game. I suppose that those childhood contacts with ticks in the Maryland fields helped prepare me for fieldwork in subtropical Florida, a haven for all kinds of biting insects and arachnids.

Conversations with wildlife colleagues, especially those from my generation, revealed that other budding biologists also had an early obsession with dinosaurs. I definitely went through a dinosaur phase. *Diplodocus, Brontosaurus, Triceratops, Stegosaurus*, and, of course, the post–Jurassic Park and now infamous *Tyrannosaurus rex* were all part of my vernacular in the early years. The plastic models that I coveted and collected were amazingly lifelike.

By the preteen years, my interest in large and dangerous extinct reptiles had waned and morphed into an obsession with large and dangerous extant African animals. While other girls in my classes wrote love notes and dreamed of prom dates and dresses, my doodling involved sketches of African wildlife, and my notebook was filled with their exotic-sounding names (wildebeest, springbok, oryx). Of course, at one point, I really thought I would be the next Jane Goodall. In fact, Jane became my hero: her groundbreaking chimpanzee studies at Gombe began when I was an impressionable nine-year-old and continued throughout my teens and beyond. At times, I was torn: Did I want to work as a veterinarian, or did I want to study wild animals in some far-flung locale? Later experience as a veterinary technician would help me decide.

My interest in critters also extended to, and overlapped with, my love of books. I have been a confirmed bibliophile for as long as I can recall. Books impart knowledge and transport us to lands of adventure and mystery. My bookshelves are overflowing, but I have always maintained that there are much worse vices than having too many books. I am so thankful that my mother and father shared and bestowed a strong appreciation for the written word. My earliest favorite books included *Home for a Bunny* and *Snow Bumble*. I don't remember much about the former book, but I still have the latter after sixty years. It's the absolutely charming story of how a British "honey-coloured Pekinese" and his mice and beetle buddies endured an extreme blizzard. The beautifully written opening words have remained in my memory all these years, with the alliterative first three words of the second sentence becoming a mantra of sorts: "For six days it snowed. *Softly, silently, ceaselessly,* the snowflakes came flurrying down. Bumble, the Macmouses, and even the Beetles were bored stiff."

My love of equines also translated to books: I adored *The Black Stallion* and the Misty of Chincoteague series. I even went to Chincoteague and Assateague Islands in Virginia on my first honeymoon to see the famous ponies. And *My Friend Flicka* was at the top of the list of my favorite books.

Books about African wildlife superseded even the beloved horse books as I got older. I would sit in my neighbor's huge, inviting oak tree and read stories about close encounters with lions and rhinos and snakes (oh my!). I was especially fascinated by black mambas and spitting cobras. Would a mamba really chase a man on horseback, and could those cobras accurately hit you in the eyes? Of course, I had no way of knowing then that snakes, specifically indigos, would figure so prominently in my future.

Especially in my teens, I was also fascinated by mythical creatures, especially the Loch Ness Monster, Yeti, and Bigfoot. I eagerly perused articles and books about them and, later in life, actually went to Willow Creek, the epicenter of Bigfoot country in Northern California. In a fun twist of fate, a Scotland tour took me cruising on the deep, dark, and delightfully mysterious waters of Loch Ness, but the mythical sea monster Nessie was a no-show.

For a high school book report, I searched my neighbor's eclectic library and chose *The Beast of the Haitian Hills*, a strange story in which an imaginary hybrid monster called a Cigouave played a role. This beast of voodoo folklore had the body of a panther and the head of a human. After the low-budget horror movie *The Legend of Boggy Creek* premiered, I even stopped by the alleged swampy haunt of the Fouke (Arkansas) Monster. And much later, when I got to Florida, I learned of the resident pungent Skunk Ape that supposedly roamed the Everglades and beyond.

The human imagination is extraordinary, and I continue to be interested in tales of Navajo skinwalkers and Latin America's Chupacabra (goat sucker). Apparently, some sightings of alleged Chupacabras in northern Mexico and the southern United States have been verified as unfortunate canids with mange. The pseudoscience field of cryptozoology was intriguing to me back in my youth, and still is, from a folklore perspective. In fact, the intertwined world of humans and animals, whether in reality or myth, would be an ever-present thread throughout my career and life.

As I look back, I can see how my interactions with animals and childhood explorations in nature helped guide me down the path toward my perhaps inevitable career choice. Moreover, I was definitely a dreamer, and the quote "Keep thou thy dreams—the tissue of all wings is woven first of them" inspired me. I had seen it in a booklet about nature and had cut it out to put up on my bulletin board when I was a preteen. Even much of my daydreaming took place outdoors: I would lie in the grass for long moments, watching the cloud shadows move across the landscape while musing about riding horseback or studying animals in exotic locales. I was also blessed with a vivid imagination and a poetic soul, which I think has been helpful in translating the sometimes staid and dry scientific world into stories and information that the layperson can embrace. I am reminded of a quote by Edward Abbey from his book *The Journey Home*: "Any scientist worth listening to must be something of a poet. He [or she] must possess the ability to communicate to the rest of us his sense of love and wonder at what his work discovers."

CHAPTER 1

Sibling Rivalry

I've always maintained that my ability to work harmoniously with men, especially in a career field that was decidedly once a male bastion, stems in part from growing up with two mischievous older brothers. My first sister-in-law often noted that it was a wonder I was as normal as I was after being subjected to such intense teasing from them. As a biologist, I know that sibling rivalry in humans and many other species is a part of nature. But my two brothers, six and eight years older than I, had definitely honed the fine art of tormenting a younger sister. Always innovative, they created a so-called tickle box, a Model T Ford ignition coil and crank that delivered a substantial shock. It took only one shock for most of their friends to avoid this implement of quasi torture. However, at a very young age, I *really* enjoyed playing the game "Ring around the Rosie," and the only way my brothers would comply was *after* I endured the tickle box.

My brothers were just as accomplished at mental games. They would stare intently at me during dinner until I would proclaim, "Why is everyone looking at me?" Dinner was considered a formal affair in our house, and such shenanigans inevitably incurred my father's ire. Because I loved horses, my brothers frequently reminded me how equines stacked up in the animal intelligence spectrum. "Stupider than a pig," they would tell me. And even worse than the stereo-pinching I endured when I had to sit between them in the back seat during long car trips from Maryland to family reunions in Michigan was being told that "Mr. Bear Squish-'Em-All-Flat" would sit on the long, dark, scary turnpike tunnels until the tunnels collapsed on unsuspecting vehicles. My dad became furious when I would scream and grab my mom's neck. He was a master of the backhand for whomever he deemed causing mischief in the rear seat. I have always loved and respected the power of bears, so the thought that this giant mythological or actual bruin was about to squash our car was terrifying indeed.

During one of those family trips to Michigan's Upper Peninsula, I shrieked when we went to see the real bears at a local dump. My child's mind thought the bear might indeed come sit on our car. And in reality, bears have been known to drastically damage vehicles that contain food. But these bears weren't interested in squashing our sedan; instead, they were partaking of the bounty that the dump offered. We now know that bears and garbage are a bad combination that can lead to increased bear-human conflicts.

If my toughness came from my brothers, my thirst for knowledge and my lifelong wanderlust came from my father. Smitty was an electrical engineer who patented a part to a nuclear submarine and brought me gifts from his own much-anticipated business travels. His innate strictness was counteracted, to some degree, by my spiritual, earth-angel mother. From her, I inherited or gleaned a strong sense of compassion and understanding. Always encouraging, Verna loved hearing about my adventures, and when my first husband, Mike, noted that I recounted stories in great detail, my mother told him that she had urged me to "expound." Sometimes, her own enthusiasm and sweet naïveté caused her to humorously mislabel. For example, New Mexico desert scrub became "scrub brushes." Later in my life, when I was an established wildlife biologist in Florida, I went down to visit my parents in New Smyrna Beach. As my mom introduced me to her minister after church, she proudly told him, "My daughter's a wild biologist." My dad didn't miss a beat: "That may be true, but she's also a wildlife biologist." My mother's mislabeling didn't imply that I was a party animal, but there was some unintended truth to my "wild child in the woods" persona. The term *wild* has many connotations, but for me, it's always been a positive term of freedom, adventure, and an existence rooted in nature.

2

A Horse, a Horse

My Kingdom (Lunch Money) for a Horse

Equine Obsession

Little girls and a passing or lifelong love of horses often go together like peanut butter and jelly. For me, the flame burned intensely and was long lasting. The country song that notes "I should have been a cowboy" (cow*girl*) resonates strongly to this day. In addition to reading every horse-related fiction and nonfiction book that I could buy or borrow, I collected amazingly realistic horse statues. Two favorites were Hobgoblin, an Appaloosa, and Gaudenzia, a white Arabian. My friends and I also ran around on our two legs, either playing that we were riding horses or that we were horses. Every Christmas list began with the same request, "a pony." To say that horses permeated my waking and slumbering dreams was an understatement. Alas, my father had other priorities besides purchasing a horse, and he wasn't going to give up precious space on our single acre for a high-maintenance equine.

Finally, my parents agreed to let me take riding lessons, but even those were unconventional. The lessons were given by a former cavalry officer, Colonel Young. My favorite horse was a bright sorrel mare, Princess Norkett, but I occasionally rode the more spirited liver chestnut gelding, Comanche. Rather than simply tell us to walk, trot, or canter, the colonel would shout, "These riders, gather your horses, trot, ho!" We sat in actual cavalry saddles, also known as McClellan saddles, named after the nineteenth-century army career officer who designed this unusual, lightweight, sturdy saddle. Characterized by a rawhide-covered open seat, a thick leather skirt, and wooden stirrups, these were not the modern, sleek English saddles or the prominent-horned, heavy western working saddles. No wonder I'm a fan of the Old West: I rode cavalry-style and grew up watching the old television westerns and John Wayne sagas set in Arizona's evocative Monument Valley.

In addition to taking riding lessons, I seized every opportunity to ride other people's horses and ponies. One of my elementary school friends, Barb, was an unrestrained child who owned a buckskin gelding named Rusty. Horses that are solidly steady and trustworthy are deemed "bombproof." Rusty was decidedly not bombproof. He was sometimes prone to fits when he seemed to enjoy dumping a rider. Combined with Barb's unruliness, he really wasn't a horse that a neophyte like me should have been riding. But he was often the so-called only game in town.

One day when I really, really wanted to go riding, Barb informed me that the only way I could ride Rusty was if I rode him bareback and sidesaddle and jumped over a small obstacle along the roadside. Of course, I knew better—but I wanted to ride so badly that I complied and, not surprisingly, fell off and landed on my right shoulder. To this day, that shoulder is turned slightly inward and is basically deformed, but I never told my parents about the incident.

Eventually, riding Rusty and taking cavalry lessons simply were not enough to satisfy my horse obsession. I wanted my own horse, and by hook or by crook, I would have my own horse. On September 23, 1963, at age twelve, I wrote, "I am going to save up my lunch money and get a horse!!!" I'm not quite sure where such a bold idea originated to deprive myself of a hot lunch, surreptitiously set aside the designated money, and supplement it with babysitting gigs.
In the spirit of full disclosure, I did not like babysitting, but I was determined to gather as much money as possible. I was literally on the threshold of my teen years and was in junior high school. Each day, my parents gave me fifty cents to purchase a school lunch. And each day, I secretly packed an apple or other snack, which I ate for lunch while watching my friends scarf down their hot lunches. My parents must have wondered why I stayed so thin when I was obviously supplementing my lunch, but they never explored the question further.

Over thirty-five weeks, I painstakingly recorded each fifty-cent increment (a whopping $2.50/week, plus any extra earned income), and I still have that handwritten record. I came very close to confiding this plan to my unadventurous neighborhood friend Bambi, which would have ended it immediately because, as a good little Catholic girl, she would have told her parents. I also almost lost the money when our house was robbed one evening while we were at a church supper. The thief trashed the house and my pink-hued room, but he must have figured the relatively small, locked cedar chest was simply filled with girly possessions of no value to him. I still have that cedar chest as a reminder of how I earned my horse.

As the school year waned, I knew I needed more money, so I bet my brothers twenty dollars each that I would have one hundred dollars by the end of the school year. They of course had no idea I was hoarding lunch money and figured there was no way "this crazy kid" could achieve that goal. My mother and one brother were present and totally shocked when I brought out the cedar chest filled with *many* quarters, other change, and dollar bills. We took the box to the bank, and when the money was counted by machine, I had $103. My brothers were truly appalled, but they paid off their bets. That left my poor mother in the unenviable position of trying to explain this to my stern father. I wisely was not present at the time of that discussion. Although Dad was angry at first that I had hoarded my lunch money and not eaten substantial meals, Mom pleaded my case and somehow convinced him that if I was this determined to have a horse, maybe they should get me a horse. I'm sure she also noted that some of my schoolmates were getting into trouble with boys and drugs; it was the 1960s, after all. Under the circumstances, a horse wasn't such a bad thing.

As the summer of 1964 wore on, it became increasingly obvious that my father was trying to stall the horse-purchasing process. I don't recall how many equines we saw during June and July, but like the proverbial love story in which all others in a room fade into the background when "the one" is observed, I became focused on a three-year-old dark bay filly. From my dad's perspective, the price was right: two hundred dollars. I was enchanted by her dappled coat and the prominent white star on her forehead. She may have technically been bay, but she was as "green" as the Maryland field grasses in summer (meaning that she was, in retrospect, probably not well trained enough for a neophyte teen). She stood just over fourteen hands, right on the cusp between a pony and horse. Like many "grade" horses, she was what I called a "Heinz 57," a mixture of breeds. Her thick neck was indicative of a Morgan, but she likely had one of the pony breeds in her as well. Her given name was Ballerina, but that did not fit her at all. I named her Flicka ("little girl" in Swedish) as a tribute to both my mother's Scandinavian background and my love of the famous Mary O'Hara novel. My imaginary equines (Boadicea) and horse statues (Hobgoblin) had been more creatively named, but somehow the name Flicka seemed to be a good fit for my long-awaited and hard-won horse.

It is often said that the initial cost of a horse is only the beginning of the financial journey. That was certainly true in Flicka's case. Because we owned only one acre of land with a house, detached cottage/garage, and large garden, there was no room at the Smith inn for a horse. Fortunately, there was a boarding stable and plant nursery up the road. Garden Gate Nursery became Flicka's new home and my daily hangout when I wasn't in school or babysitting to help support my new acquisition.

Boarding Flicka at a stable with other equines came with an additional cost beyond the twenty dollars per month in summer and sixty dollars in winter: occasional hefty vet bills. As the new filly in the pasture, she initially had to endure biting and kicking from the more dominant mares. Thankfully, most injuries were relatively superficial and could be treated with Absorbine or gentian violet. At times, Flicka looked almost psychedelic with purple-hued smears on various parts of her body.

But a more serious injury occurred early in her stabling at Garden Gate. As I was returning Flicka to her designated pasture one evening, the more troublesome mares charged toward her as soon as she was turned loose. In her attempt to flee, she ran into the barbed-wire fence, which (to my horror) bent and twisted around her. Ironically, a similar fate befell the literary Flicka. My horse sustained lacerations that required medical attention, but unlike the fictional filly, her cuts thankfully did not become infected. While I've learned that barbed wire is not desirable fencing for equines, at the time, I was at the mercy of the stable owners, who had a vast acreage to fence and were also raising cattle. I still have a scar above my right knee, a reminder of being pressed by the cattle while trying to negotiate that sharp wire fence.

The Garden Gate stable had once been a dairy barn and had long troughs that ran the length of the building. That unusual design turned out to be a good thing because my new equine-child did have one very bad habit. Each stall had a spigot in it, which made watering the horses much easier. But Flicka decided she didn't want to wait for a human to turn on the water and fill her bucket: she would use her teeth to turn the round handle until she got a good, fresh stream going. None of us, even the patriarch stable owner, was sufficiently strong to turn that handle tight enough, and one night, Flicka would have flooded the entire barn if it weren't for those troughs. The water had flowed out of her bucket, into her stall, and then, thankfully, into the hallway trough. The barn was still quite wet, and she temporarily became "equine non gratis" until someone came up with an idea to build a wooden box over her spigot.

Horses are, in some ways, as fragile as they are beautiful. Those slender legs that bestow such incredible grace also are prone to myriad afflictions and injuries. Additionally, horses cannot regurgitate and suffer from a number of digestive disorders. A primary cause of death is the dreaded colic, where gas can cause extreme abdominal pain. One winter day, I received a call from the stable owners. Flicka had eaten the same frozen alfalfa in the pasture as all the other horses, but she alone was ill. I'm sure I ran most of the half-mile shortcut through forests and fields up to the stable. The vet arrived and put a tube into Flicka's stomach to apply what was likely an oil to help her defecate. Now it was up to me to keep her from rolling and thrashing and literally busting a gut. As the late afternoon turned to evening, I walked her up and down the narrow barn hallway, where it would be more difficult for her to roll. If she did lie down, I was instructed to tug on her halter, yell, and even use a crop if necessary to make her rise again. It was both stressful and heartbreaking. This profound early lesson in responsibility continued to resonate when I was a field biologist tasked with the welfare of imperiled species.

Eventually, the stable attendant and other horse owners left for the day, and I faced this challenge alone. Shortly after dark, my father (who was adamant that I not be out after sunset) came up to the stable and told me it was time to come home. Rarely did anyone stand up to my father, but that evening, I told him about this responsibility I had and that there was no way I was coming home until I was sure my horse was beyond the critical stage. He left, and I remained in the cold, dimly lit barn with only my sick horse and the other equine inhabitants for company. I truly don't recall when I did get home that night and how I managed to go to school the next day, but the only thing that mattered was that my much-loved, equine-kid had survived.

Fortunately, my four years with Flicka were defined by much more than vet bills and close calls. Riding her was my ticket to adventure as well as an outlet for teenage angst and restlessness. As I galloped across the rolling, verdant hills of Garden Gate, I imagined that I was either Zebulon Pike (a nod to Wild West history) or, if riding bareback, that I was part of an Indian tribe. My friends and I sometimes swam our horses in the lake that fronted the stately white mansion of Garden Gate's owners. It's such a rush to feel the power of a horse swimming under you. I felt free and, yes, in some ways, wild.

It is not an understatement to say that Flicka and I basically grew up together. We evolved and matured as time went by, and she became a well-trained pleasure to ride. The first year, however, it was not uncommon for me to ride Flicka over to my homestead, only to have her buck me off in front of my horrified mother. She really was a sweet filly, but one time when she was "in season," she turned in her stall, laid back her ears, and sunk her teeth into my abdomen. I would have undoubtedly been injured if I hadn't had on a thick down coat. Even with the coat, my bruised stomach went through a rainbow of colors. Needless to say, she was seriously reprimanded and never attempted such a vicious bite again. The only time she kicked was when she was injured and I was applying Absorbine. In one case, the stinging fluid went directly into my eye, forcing me to quickly morph from horse-nursing to eye-flushing mode. Despite these rare transgressions on her part, I absolutely adored my horse and would often gently grab her chin and plant a kiss on her velvety nose.

My Brush with the Law

I have discovered that I am drawn to colorful characters. This recognition of, and fascination with, amusing and unusual humans became a thread that wound throughout my life and career. My best friend at the stable was Becky, a funny and delightfully quirky girl my age. Becky and I had many misadventures during those horse-dominated years. We dubbed ourselves "Otis" and "Jake" (I think I was Otis) and pretended we were cowboys riding the range. On one occasion, we somehow got the crazy idea that both of us should ride Tiny (a decidedly misnamed, giant draft horse mixed breed) bareback out to his pasture with only a halter and lead rope to guide him. Of course, something spooked the massive gelding, and he charged toward the gate of his pasture. Becky was sitting behind me and was screaming as the fence loomed closer. Before Tiny could decide whether to shy away from the gate or try to jump it, Becky pushed me and we both crashed to the ground. Luckily, Tiny's huge hooves missed us, and he simply stopped at his pasture entrance and looked back at us. The stable manager, Bob, walked slowly over to us and dryly asked if we were trying out for some Wild West show.

Both Becky and I were good kids in an era of sex, drugs, and rock 'n' roll. However, we did have a single brush with the law. One of the other stable gals, Barbara, was older and more mischievous. At the time, her parents were overseas, and she was staying at the stable owners' home temporarily. Rolling houses with toilet paper was all the rage, and somehow, Barbara convinced Becky and me to decorate our friend Sherry's house. I knew better but got caught up in the fun. Unbeknown to us that evening, Sherry's brother apparently saw Becky's car pull away from the scene. The next day, we three amigas returned from an enjoyable trail ride to find a police car parked outside the big white mansion on the hill (never a good sign). The matriarch owner of Garden Gate came out and told us to put our horses in the pasture and come immediately up to the house. We were busted.

She and the cop confronted us in the elegant living room. We of course confessed, and Becky began to cry. I was contrite while Barbara remained quasi-defiant. I was told by the stable owner that she expected this from Barbara, and maybe even Becky, but not me. I've always been blessed or cursed with being the responsible, commonsense one. She also told me that she wouldn't tell my father because he frequently threatened to sell Flicka when unsavory events occurred. The cop was obviously amused by the three "perps" before him, and he asked why we weren't out on dates like other girls our age. Becky tearfully replied, "Because we're *rejects*." With that, the cop could no longer contain his amusement. He shook his head, laughed, and we were reprimanded and set free.

First Transition: College Bound

Time waits for no man or teenage girl, and, in 1968, my four years with Flicka were drawing to a close. I had been accepted to Murray State University in western Kentucky and would be leaving that summer to keep my sister-in-law company at nearby Fort Campbell while my brother was in Vietnam. Additionally, I had grown four inches in one year while Flicka remained the same height. I now looked a bit tall in the saddle. Not unexpectedly, selling my horse was truly one of the hardest things I ever had to do. Alas, this transition was part of growing up and moving onward. I found a ranch where Flicka would be welcomed and appreciated. It was, however, a bit disconcerting when one of the resident stallions almost kicked down his stall as Flicka's transport arrived. I don't recall the details of what must have been a heartbreaking farewell, but afterward, I was glad that I could leave my horse-haunted Maryland homestead and embark on adventures of a decidedly different kind.

I had fulfilled my earliest quest: to have my own horse. My world had revolved around equines; the four years had been both a learning experience and adventurous beyond my dreams. But now I had a new and different desire: to compile a résumé of training and experience so that I could someday find a fulfilling job working with animals.

3

Would Dog Bite (Mail)Woman?

Deer in the Headlights

I must confess that my undergrad years were not necessarily memorable. I had come from a premier high school into college classes where, sadly, many students struggled with English grammar and syntax. I was often bored in the required, basic classes but found stimulation and inspiration in my declared major of biology. At the time, I jokingly referred to biologists as "God's chosen people" because we studied life. I was fortunate to get a position as a lab assistant in botany and was allowed to actually teach the labs and prep the tests for the female professor, fondly known as Doc. This was unusual because I was only a sophomore when I started teaching labs, and there were older students and even occasional grad students in the botany labs over the next two years.

My affinity and appreciation for plants continued throughout my wildlife career. Knowing the plant species that serve as forage or that otherwise affect animals positively or negatively is paramount to managing wildlife habitats. And I guess that I've always been a flower child because I am strongly drawn to colorful floral blooms.

In addition to the various biology and other required classes, I elected to take an introductory art class. I had adored art in high school, where my efforts were praised by my teachers. Of course, many of my drawings and paintings were of animals, especially horses and giraffes. But the college course was a letdown because we were expected to draw boxes and human figures. (How boring is *that*?) I understood that we needed to learn the basics, but it just wasn't my cup of tea. The teacher was not terribly amused when I turned in a watercolor of a rhino knocking over a field vehicle; my grade reflected his dissatisfaction.

My first college roommate was an art major, and she thought this creative but rebellious painting was hilarious. The idea for this piece likely stemmed from one of my favorite movies, *Hatari*, where John Wayne captures African animals for zoos. Although my overall grade in that class was the lowest of my academic pursuits (a C, heaven forbid; I was an A student), I didn't let it dim my appreciation of art. Today, I express my artistic creativity through photography, and I regularly frequent art shows and galleries. And my home is filled with wildlife art.

I also took a college course in creative writing, another love. I had been fortunate to have an amazing English teacher in high school, who encouraged me both in the written and spoken word. She especially got a kick out of my oral book report on Owen Wister's iconic western novel *The Virginian*: "When you call me that, *smile*." And she revealed my flair for the dramatic by casting me as Beneatha, the rebellious (see a trend here?) sister of African American Walter Younger in *A Raisin in the Sun* (presented to all the English classes). Those of us with pale skin donned darker-hued makeup. In fact, I slapped the guy that played Walter so hard during the play that some of his makeup came off on my hand.

My culminating piece for the writing class was a short story set in Africa called "The Man Killer" about a rogue cheetah. I still have that story, but I now know that if I ever revise and publish it, the big cat would need to be a lion or leopard. Cheetahs are blazingly fast but basically docile felines that rarely attack humans, and the relatively few attacks that have occurred involved captive cheetahs. Actually, this story was a play on words—because its theme was how humans destroy wild animals and habitats. My ardent conservationist side was emerging.

But one undergrad class truly lit my fire: wildlife management. This class, taught by a highly regarded wildlife biologist from the Tennessee Valley Authority, initiated the spark that would later fully ignite and lead me to my grad school research and eventual career. For one of our field trips, we went out to Land Between the Lakes, a scenic national recreation area between Lake Barkley and Kentucky Lake. We were there to help trap and survey fallow deer, a Eurasian species that has been introduced into many other parts of the world. The population at Land Between the Lakes was introduced in the early twentieth century and is likely the oldest established population of this species in the United States.

The details of this long-ago field trip are a bit hazy, but if memory serves me correctly, the deer were captured using a cannon net. As the name implies, projectiles propel a large net over birds or mammals drawn into the targeted site by a bait. What I strongly recall is the moment that we released the deer after checking them over and gathering data that the wildlife biologists needed. We had retreated to the trucks and were waiting to see what the deer would do. By now, darkness had descended. Much to my surprise at the time, instead of running back into the safety of the forest, several deer spooked and ran directly at the lights of the trucks. In one case, a deer jumped up on the hood, bounced off, and dashed unhurt around the truck into the trees. Although most "deer in the headlights" cases involve light blindness that causes deer to freeze in place on roads, this was a memorable case of the light drawing the deer like a moth to a flame.

Mailbag Blues

Throughout my career, I never purposely targeted a male-dominated job just to prove a point or serve as some sort of pioneer. I followed my interests and my heart, and sometimes atypical jobs came my way. Summer jobs for college students were notoriously difficult to find during the late 1960s and early 1970s, especially in an arena where I was competing with University of Maryland students. I needed to work, so I applied to the 1970 Washington Summer Internship Program. Because I was a biology major, I had hoped to get a job in wildlife management or animal husbandry. Alas, there was no room at those coveted inns. Instead, I was informed of an opening in the US Postal Service (USPS) as a summer substitute mail carrier just outside D.C. And so began an arduous, humorous, and unforgettable summer adventure in the Maryland suburbs.

At that time, summer substitutes were given only one day of training, generally worked six days a week, and could anticipate a different mail route each day. My male coworkers did not expect me to last the summer, and I think they gave me the most challenging routes on purpose during my first week. Despite much teasing, a teamwork ethic was evident: After finishing their own routes, three of the guys came to assist me on my first day, a terrifying day because I had to drive a right-hand-drive Jeep with no prior experience or instruction. Traffic in the Maryland suburbs was daunting in a regular vehicle; the altered perspective of a right-hand-drive Jeep only exacerbated the fear. I later found out that I was supposed to have a special driving license and that my supervisor got into trouble for sending me out sans training or license.

Obviously, my training (or lack thereof) did not cover the necessity of the USPS unofficial motto: "Neither snow nor rain nor heat nor gloom of night stays these couriers from the swift completion of their appointed rounds." I knew I had a huge responsibility to get the mail delivered, but I was faced with a dilemma when the first torrential rainstorm descended. I stood on a porch wondering how to complete my route without getting the mail wet. My supervisor was not pleased. After that, the mail, although sometimes drenched, always went through.

The long hours, intense summer heat and humidity, and physical work were certainly good preparation for biological field research. Rising at oh-dark-thirty, I hurried to the post office to case the mail (put it in alphabetical order) before bundling it and actually beginning my route around midmorning. I then drove to a specified area, parked the Jeep, placed a relay of mail in the leather bag, and walked to each mailbox delivering that relay. After multiple relays and much blister-inducing walking, I returned to the post office sometime in the afternoon. I often went home with print all over me, and I literally wore holes in my new leather shoes after six weeks. And while the regular mail carriers wore uniforms, I wore jeans and a light-colored shirt, leading to numerous cases of mistaken identity. Kids thought I was a hippie; women thought I was a feminist and seemed to support a mail*woman*; and some men just thought I was either crazy or was attempting to steal the mail when I picked it up at the boxes. In one instance, a local cop looked me over and commented, "You gotta be kidding!" It is almost incomprehensible in this modern gender-bending, role-reversing, and equal-opportunity era that people could be so naïve and reactive in 1970 about a girl simply delivering the mail.

As charted throughout my unconventional career path, a recurrent theme became apparent in job fields traditionally populated by men: the somewhat sexist (yet admittedly successful) use of a woman to enter scenarios where, in the same circumstance, a man would engender suspicion, hostility, or even downright fear. My post office supervisor was apparently no stranger to what a later wildlife boss somewhat tactlessly declared, "We got to use our women." Absurdly early one Saturday morning (I'm talking 7:00 a.m.), I was dispatched to an apartment complex to deliver packages. The inhabitants didn't get nearly as angry when they opened their doors to a smiling young female face.

But it wasn't the public-perception issues or British-style steering wheel placement or intense thunderstorms that presented the greatest problem: It was man's best friend. I adore dogs and generally get along fabulously with them. However, it's an entirely different story when you are entering their yard carrying a mailbag. Perhaps they feel threatened or think that you're about to take something. Whatever the reason, canines can be a major challenge for the walking mail carrier. One of my tall, strapping, colleagues, Otto, was bitten so many times that our supervisor threatened to fire him if he was bitten again, a threat that extended to the rest of us as well.

Why would a supervisor issue such a mandate? The reason is that mail carriers, regular or substitute, are issued a special spray to repel and subdue attacking dogs. Obviously, Otto either didn't use his spray or didn't have a fast-enough draw. I tried not to use the spray, even though it doesn't apparently harm the dog. I attempted an alternative, somewhat successful tactic: talking the agitated or aggressive dog out of biting. Because this College Park, Maryland, suburb was hardly free from crime, there were some quite vicious dogs whose primary purpose was guarding their turf. I definitely had close calls and almost got bitten on several fear-laced occasions.

Of my ten or so rotating routes, one had notoriously aggressive canines of all shapes and sizes. At times, it was almost comical: I'd get on the porch of the house, and then three shrieking little dogs would converge on me. In those cases, one had to have the quick draw of a Wild West gunslinger. I soon became an expert on dog barks and behavior and typically could rate the relative threat. However, in what could have been a fateful day, my guardian angels imbued me with a quick reaction. I had entered a yard and heard loud, aggressive barking coming from a screened porch. Suddenly, the screen door crashed to the ground and a large, truly scary dog of undetermined breeding charged toward me. The rarely used spray was on my mailbag, and amazingly, I swiftly employed it directly into the dog's gaping mouth and angry eyes. Shocked, he stopped, blinked, seemed disoriented, and went over to the flower garden to pee. The equally shocked owner abruptly and belatedly appeared on the scene, and I apologized, delivered the mail, and hurried away before the dog or human could rally. I truly felt terrible at the time, spraying a repellent into the eyes of someone's beloved pet. But the survival instinct fortunately overrode my innate love of canines.

Second Transition: Marriage and New Mexico

My undergrad training and summer jobs were soon behind me, and both Maryland and Kentucky were about to be in the proverbial rearview mirror. I was armed with knowledge and had proven myself tough enough to toil in the heat and humidity. To say that I was ready to set the world on fire is an understatement. As with any young college grad, the unknown trail loomed ahead, both exciting and daunting. And now I had a new, even more important quest. I wanted to actually begin my life's work with animals. At that point, I didn't know exactly how I would benefit the beasts, but the flame was there and burning brightly.

When I graduated from Murray State University with a BS in Biology in 1972, I naïvely thought that I would realize my earlier dream to become the next Jane Goodall. It didn't seem to bother me that I had never been to Africa or had any training as a primatologist. There were plenty of other species to study besides chimps. I also ignored the abysmal statistics about new grads actually finding a job in the biological field. To further thwart my African dreams, I fell in love with a fellow dreamer whose life goals did not necessarily match mine. When I became engaged to the tall, handsome man-child that I had met in college, my dad expressed some concerns about the future. Dad liked Mike, but he could see what I couldn't, that Mike would have trouble accepting my career path.

Of course, I didn't listen; I was in love. And my sweet mom and her church ladies were totally charmed by my fiancé. Mike had skipped a semester to help his family up in Louisville and was quickly drafted. To avoid the army, he joined the air force as a dental technician, and we wrote each other nearly every day while I was finishing my degree. But love letters, poems, and sporadic phone calls don't equal quality time together in the same location, and when we did marry in the summer of 1972, we had a lot to learn about each other and marriage. We piled all our belongings into Mike's Volkswagen Bug and dodged tornados as we made our way west from Fairland, Maryland, to Albuquerque, New Mexico. Elton John's "Rocket Man" was *the* song that summer and heralded our arrival in the Land of Enchantment as we settled first into a tiny apartment and then into a house on Kirtland Air Force Base. Little did I know that the marriage would not last, but my love of this strange, seemingly sere, landscape would only grow stronger with time.

4
Going Nuclear at the Veterans Administration

Job Seeking

Despite the dearth of jobs related to biology in Albuquerque, I was nevertheless determined to work with animals. My first position postbaccalaureate was indeed humbling: I cleaned cages and waited on customers at a small mom-and-pop pet store near the military base. The dogs and cats at the pet store were not a problem for me. It was the birds that proved scary. The parrots especially had a tendency to bite the hand that fed them. And the mynahs were loud and raucous; one seemed to be recounting the fate of his amigos as he repeatedly screamed: "Six and Seven are sold!"

Early in my initial New Mexico tenure, I was sometimes forced to find better-paying work that didn't necessarily further my goals to do field research and work with animals; these were side trails for sure. An enlisted man's salary just wasn't keeping up with our bills. Moreover, Mike really liked to shop and would spontaneously buy items large and small that we didn't need. So I took jobs as a cocktail waitress, gift shop clerk, and substitute teacher. Regarding the latter job, the colorful wife of one of the air force dentists noted that she would rather be a prostitute than substitute teach. I understood her point when a brazen student went through my purse and claimed that I was too poor to rob. Luckily, I also found work that did relate to my future career field in one way or another: nuclear medicine research technician, veterinary technician, and helitack firefighter.

Rats, Rabbits, and Monkeys

Interestingly, and perhaps ironically, I may not have been hired as a senior biologist by the state of Florida in 1980 if I had not had prior published research experience, albeit in a totally unrelated field. That experience came via a research grant in 1973 from the University of New Mexico School of Medicine at the Veterans Administration (VA) Hospital in Albuquerque. Although far from the realm of wildlife biology, these laboratory studies were indeed scientific research. I had been hired as a technician in nuclear medicine and was excited about participating in investigations linked to finding an early detection for arteriosclerosis (hardening of the arteries). More specifically, we were looking at a form of arteriosclerosis called atherosclerosis, which is caused by the buildup of fatty plaques and cholesterol in the artery walls. No, we were not making humans glow in the dark. The research at that time involved looking at a radioactive lipophilic (fat-loving) dye, and the studies involved rats and rabbits.

As with much medical research (and wildlife research, for that matter), such undertakings are not for the faint of heart. In a nutshell, the rat portion of the research was basically injecting the lab rats with the radioactive dye and determining its distribution in selected organs postmortem. If handling rat parts was more than a wee bit unsavory, working with the rabbits had its share of challenges too. It was part of my job to feed cholesterol to a subset of rabbits I would mix the slightly sweet-smelling cholesterol into their pellets, and over time, their ears and eyes showed signs of fat deposits. Seeing such overt evidence of the cholesterol should have been enough to deter me from ever eating a cheeseburger again. We would then surgically examine the aorta (main artery) to determine differences in the anesthetized control and cholesterol-fed rabbits regarding how this radioactive, fat-loving dye (which collected in the fatty plaques) behaved postinjection.

I must confess to being torn about this research. Although these lab animals were raised for this express purpose, my critter-loving side felt bad for them, especially when our janitor vociferously protested their use in our research. But my scientist side tried to rationalize (not always successfully) that the findings from these studies could, and likely would, save many human lives. What if someone I loved needed this early detection and could benefit from it?

I was at least grateful that the radioactive isotope we were using, Iodine-125, has a short half-life. One fateful day, I was working under the chemistry hood when, instead of the fan sucking air, it blew radioactive material into my hair. I was forced to stay at the lab for many hours after work, washing and rewashing my hair with a decontaminant before I could go home. Some of my friends later claimed that this incident may explain a lot about my hair, which has always been a little bit wild.

Occasionally, I was also called on to work with the VA's research monkeys (way too close to humans). A different radioactive isotope was being injected into a vessel in the groin to determine how it moved through the body. I'm not sure why we named them because that only made the research undertakings more poignant. Alex and Thor were stub-tailed macaques, and I must confess to feeling quite bad for their plight. They were decidedly, and understandably, not friendly—and would show their impressive front teeth in so-called canine displays that made them look like they were yawning. I was glad that my tenure was over before they had to be sacrificed, especially since I had once kept Alex alive using an ambu (resuscitator) bag to breathe for him until he came out from under excessive anesthesia.

My only other contact with monkeys involved yet another fateful day when several of the lab freezers had malfunctioned overnight and I arrived to an odor beyond description. The sacrificed monkeys from long ago had thawed, and as a low woman on the totem pole, I was a major player in the absolutely disgusting cleanup effort. Neither masks nor periodic gasps of outdoor air made the stench more bearable, and I wondered at the time whether this job really would help my incipient résumé. I later learned that distastefully odoriferous animals, both alive and dead, are a given in wildlife biology.

The Quick and the Dead

Yet another strange job responsibility was related to radiation safety: If a VA patient had passed away after undergoing any type of treatment or diagnosis involving radiation, the radiation safety officer (my supervisor) had to go down to the morgue in the bowels of the building to check the level of radiation in the deceased person. Of course, as a nuclear medicine technician, I was expected to assist. To say that this was my least favorite undertaking (pun intended) is an understatement. *Creepy* doesn't even begin to describe the locale and the scenario. Fortunately, this was a relatively rare event; nevertheless, I had to steel myself and retain my professional demeanor.

My time at the VA coincided with the post–Vietnam War era, and, sadly, the psychiatric ward was filled with soldiers suffering from PTSD before posttraumatic stress disorder was officially described or understood. I would encounter some of the more functional patients when I went down to the second floor to copy scientific papers for my supervisor. Nuclear medicine was on the fourth (top) floor at the time. As a medical technician, even one in research, I was expected to wear a uniform that identified me as an employee. Early one morning, I got on the elevator as I did every morning. The elevator uncharacteristically stopped at the second floor, and a somewhat wild-eyed, young male patient got on. It was immediately obvious that this was not a patient who should be wandering around the VA. As he stared at me, I tried not to become self-conscious or to show fear. The elevator seemed to move agonizingly slowly. Finally, he spoke and asked, "Can I hold your thumb?" Considering that he could have asked to hold other body parts that would prove considerably more threatening, I agreed. And so we rode up to the top floor with my thumb in his grip. I gently extracted my thumb as the elevator door opened, bid my companion adieu, and as soon as the elevator door closed, summoned the medical cavalry. As far as I know, the young man was safely returned to the psych ward.

I'm Not Your Pinup Girl

As if dissected rats, filleted rabbits, gag-inducing thawed monkeys, radioactive corpses, and PTSD veterans weren't challenging enough, I had to deal with an extremely thorny issue that I had never encountered before: *sexual harassment*. Actually, the term as we know it today wasn't even in use back then. It was apparently being discussed in a few academic and activist circles during the 1970s but wasn't more widely known until the early 1990s during the Anita Hill/Clarence Thomas testimony. In 2017, the lid was totally blown off that sealed and secretive Pandora's Box, and the demons of sexual abuse, harassment, rape, and other mistreatment of women (and sometimes men) in the workplace came forth. I don't tweet, but if I did, I could have officially joined the #MeToo movement. Although not without controversy, this movement was a pivotal moment in the history of women's rights and gender equality. Only when such disgusting behavior is out in the open can we begin to fix the problem and mandate a safe working environment for all.

As a young professional in the early stages of my career, I didn't want to make waves or get fired—but the perpetrator was my immediate supervisor. To add to my confusion on how best to handle this escalating, unpleasant situation, Dr. H (as we'll call him; the Dr. being a PhD) wasn't the stereotypic, masher dude ("Hey, baby . . ."). He was more analogous to Peter Sellers's portrayal of the bumbling detective in the Pink Panther movies. I suppose it could be said that this balding, middle-aged, married man developed a crush, but it was more calculating and over the line than a mere crush. He told me that I resembled a Peruvian princess. (Note: I have the blond hair and blue eyes of a Scandinavian, not the dark, exotic features of a modern-day Inca.) His ancestors hailed from South America, although he proclaimed himself "pure Spanish."

One of my first clues that a problem existed occurred when I was working in the darkroom developing film. Dr. H actually came into the darkroom and ineptly tried to kiss me. I was truly shocked, and he became embarrassed and departed. On another occasion, everyone had to be checked over for radiation after a minor scare, and Dr. H made sure that the wand of the Geiger counter tarried at my bosom. Another really weird scenario involved a photo of me in a wetsuit after a trip with Albuquerque's Desert Divers to the Sea of Cortez in Mexico. Dr. H had the audacity to surreptitiously steal the photo from my desk and pin it up on his office bulletin board.

At the time, my young husband, Mike, who towered over my

boss, wanted to come and "beat the tar out of him." As tempting as that seemed, I didn't think it would help my career. So I tried to talk to Dr. H, who always became embarrassed and apologetic—until the next incident. Perhaps the proverbial last straw happened when he and I had to travel to Las Vegas for a professional conference. As I used the hotel key to open the door of what I thought was my room, he followed me in and confessed he had only reserved one room. I was shocked and absolutely furious. Thankfully, I had gal pals in Vegas and called them to rescue me. Again, I tried to address this absurdity in a calm, direct, professional manner when we got back to Albuquerque.

Finally, I told several of the other technicians of this situation because, as we've seen, such horrid behavior thrives in secrecy. I think Dr. H was now aware that others knew, so he retaliated by curtailing my leave time and began treating the others strangely as well. Enough was enough: I went to the head of Nuclear Medicine, a quiet, Anglo medical doctor, and told him only of the mistreatment regarding leave. I alluded to other problems, but I was too young and embarrassed to try to explain something that had no definition yet. He sadly laid out the reality for me: "I'm sorry, but he's Hispanic in a Hispanic town, and I can't do anything."

Fortunately, the grant that was supporting me and the other techs was drawing to a close, and we all quit early as a united group against this untenable situation. Whatever message was relayed to the head of Nuclear Medicine, or our supervisor, I will never know. I left the VA and didn't look back. However, if that had happened later in my career, I would have gone the full mile to terminate the illegal behavior and to make sure no one else had to endure it.

I was once again in need of a job, and as bizarre as the nuclear medicine stint was, it was nevertheless an important stepping-stone to my eventual wildlife research position in Florida. I was now about to enter the realm of veterinary medicine, to see if I wanted to pursue vet school or go another direction. I will always be thankful for the path that I took. But to find and follow that path, I would first undertake two more positions in New Mexico that were as different as night and day.

5
Canines and Felines and . . . Willie Nelson?

Wrangling Pets and a Temperamental Vet

Although I was not formally trained as a veterinary technician, my background in biology and medical research experience, combined with my poop-scooping skills acquired at the pet shop, apparently persuaded Dr. Ned to hire me at Academy Animal Clinic. To say that Ned was a frequently volatile, colorful character and a stern taskmaster boss is an understatement. He was truly an excellent clinician, but his terse bedside manner, anal-retentive attention to detail, and behind-the-scenes temper made working for him both a challenge and a character-building experience. He was a short, nice-looking man in his thirties who had recently opened his own practice in Albuquerque. He needed a Jack (or in this case, Jill) of all trades: I was the kennel cleaner, bookkeeper, receptionist, dental technician, exam room assistant, and surgery assistant/anesthesiologist. At times, the job could be overwhelming. And Ned insisted that I dress up for the job: no jeans or scrubs allowed. One of our well-known patients was a large male Doberman who invariably peed on my leg, so I made sure that I wore a green pantsuit on days when we examined him.

Although Ned could be testy and confrontational during everyday clinic happenings, it was during surgery that his temper reached its zenith. I frequently had to get him a fresh pair of gloves after he had put his fist through the wall when something in the surgical process went awry. Fortunately, Ned was a huge fan of Willie Nelson—and playing Willie tapes helped calm him down considerably. I'm not sure the walls could have survived without Willie's smooth voice.

On one memorable (and slightly scary) occasion, I was doing the books late in the day when two young men came into the clinic. They proceeded to flirt, ask inane questions, and generally act like goofy young males. It so happened that Ned had just acquired a long-barrel pistol that looked like it belonged to a gunslinger from the Wild West. I truly felt that the guys posed no threat and that I had the situation well in hand. Apparently, Ned disagreed; he had been in his office listening and suddenly came around the corner with that imposing gun drawn. Time froze, and for a moment, those clueless dudes looked like bug-eyed cartoon characters before they darted out of the clinic.

Ned's surgical skills and patience were both put to the test one fateful Friday afternoon. A large, mixed-breed dog that had been hit by a car and was dealing with a paralyzed leg decided it was time to take matters into his own teeth. So while his owner was at work, the dog began to amputate the forelimb. Canine teeth are not the most efficient way to remove a useless limb, and the horrified owner rushed the unfortunate dog to our clinic just before closing time. Because the unused leg had atrophied over time and had been considerably gnawed upon, trying to remove it became even more challenging. I'm sure I must have given Ned several new pairs of surgical gloves as we struggled to finish the amputation and save the dog's life. At one point, copious amounts of blood spurted onto us and one of the holey walls because the vessels were not necessarily where they would have been on a healthy, unchewed leg. The operating room now looked like a murder scene from a B horror movie. We were in surgery for many hours that night but at last completed our complicated task. I was relieved and heartened to hear that when Ned checked on the patient early the next morning, the dog was standing on his three healthy legs and wagging his grateful tail.

Most of the canines and felines that came into the clinic ranged from resigned to somewhat scared to truly terrified. A few, however, were notably aggressive. Lucifer (the name alone should have been a clue) was a massive Great Dane with attitude, and the chaos that ensued in the exam room during his visits was awesome to behold. Obviously, he had to be muzzled or someone would have lost a body part. Even more frightening was Hans, a beautiful but totally psycho German shepherd, owned by a couple who knew he was dangerous to others but who loved him dearly. Because Ned was friends with Hans's owners, he begrudgingly agreed to kennel the dog when they were out of town; undoubtedly, no other clinic or boarding facility would have done so. If we needed to address any of Hans's physical issues during that time, we had to each climb up on one side of the kennel run and dual lasso him before we could muzzle or tranquilize him.

Ned knew this dog was both extremely vicious and unhappy, and he tried to talk to the owners about potential liability. Hans's saga came abruptly to an end when he attempted to attack the couple's newborn baby. Ned and I had worried about this possibility as we watched the owner's belly grow larger during her much-anticipated first pregnancy. While not surprised, I was extremely saddened when I came through the clinic's back door one morning to find Hans's body in a bag; the couple felt compelled to have him euthanized the night before. Heartbreakingly, dogs like Hans are ticking time bombs, and he was just too aggressive and unpredictable to be rehomed elsewhere.

Although aggressive dogs could certainly inflict greater bodily harm, cats were sometimes almost as scary despite their smaller size. I was never bitten by a dog while working for Ned, but a cat's sharp tooth did pierce my thumbnail. And Pedro (or was it Pablo?) was in a class all by himself. This huge cat came to our clinic with some frequency because of urinary tract issues. Only his owner could put him into a clinic cage. To catheterize him, we had to first squirt a liquid tranquilizer into his perennially gaping and hissing mouth. Even walking by his cage was risky because he would rapidly extend his long foreleg and attempt to scratch anyone within reach. I sometimes wondered if he was just an oversized domestic cat or some strange cross between a large-breed cat and a species of wild feline.

Snakes and snails are more than fine, and wagging puppy dog tails are incredibly endearing, but I was not a fan of cutting off puppy dog tails. Bobbing tails and removing dew claws are generally done on newborns of certain breeds, and as the person holding the squealing wee pups, I was less than enchanted. In fact, I absolutely dreaded those procedures at the time, but I suppose it was solid experience for wildlife research where biologists sometimes have to inflict momentary pain to complete a necessary task. Although ear crops are done under anesthetic on older pups, that procedure was also stressful because Ned had to make sure the ears were identical and cosmetically pleasing (not too short or long or oddly shaped). After the surgery, it was the owner's responsibility to keep the splints on for enough time to render the ears upright. God forbid if a Dobie or Dane turned out lop-eared. I can't imagine being a cosmetic surgeon for vain humans.

As Ned's practice grew over time, it became evident that more hands were needed. Additionally, I was eventually going to be leaving to pursue graduate studies. Actually, I had left Ned on one occasion to pursue other opportunities (firefighting), but financial needs later forced me to return. Now there were two of us to deal with Ned. I found that I enjoyed having an outspoken compatriot, Barb, to commiserate with about Ned's eccentricities. In the early days of working for Ned, I was understandably intimidated. But later, I would just roll my eyes or give him the finger (out of his eyesight, of course) over his sometimes unfair demands. When I really wanted to aggravate him, I would simply walk up behind him and reach over his head as he tried to grasp some item on a top shelf. Barb couldn't resist telling him about the finger, so it's a wonder that I didn't get fired. But I knew Ned liked and respected me despite his frequent outbursts directed at whoever was in close proximity.

An Epiphany

Working at Academy Animal Clinic gave me the insight that I needed: Veterinary medicine was an admirable and fascinating career field, but I came to the conclusion that clinical work was too confining for me. Moreover, a vet treats the pet owners as much as the animals themselves. I was to later learn that there is no escape from dealing with human psychology, and wildlife biologists can rarely sequester themselves away in the woods for long. In 1977, the siren call of the fields and forests resurfaced, and I would soon search for a fitting graduate program in wildlife biology.

My imminent departure meant that Ned had to hire my replacement, and I was allowed to weigh in on the choice. There are many criteria for a good vet tech, but I always felt that the "I love animals and really want to be able to pet them all day" sentiment wasn't the ideal determining factor. Loving animals is a fine trait and a useful one for someone working in a vet clinic. However, most of the animals are scared, and while reassuring them is a good thing, too much softheartedness can compromise effectiveness. Working with animals in a clinical or research scenario requires equal measures of compassion and toughness. Rhonda had those desirable traits, plus she was a quick learner, efficient, and quite personable. She was also, by chance, drop-dead gorgeous. That latter characteristic didn't matter to me, of course. I wholeheartedly gave Rhonda my vote as my successor because of her qualifications. But I had to laugh when I later learned that Ned had divorced his wife (trust me, I did her a favor) and married Rhonda.

Years later, I went back to visit Ned and found a very different person, much calmer and more accepting of life's challenges. Maybe Rhonda had helped, but more likely it was the result of a devastating auto accident in which he ran his Alfa Romeo convertible under a semi and had somehow survived. A portion of his brain had actually come out of his skull, and it seemed when they put him back together again, a new and more accepting Ned emerged. Perhaps he was like the stinging caterpillar that had morphed into a butterfly.

6
Whirlybirds and Conflagrations

My Summer as a Helitack Firefighter

A Long Commute

With experience under my belt in both nuclear medicine laboratory research and clinical veterinary medicine, I nevertheless realized that I still needed field experience if I wanted a career in wildlife biology. Yet I was as surprised as anyone when my federal application for zoology or wildlife positions landed me a job as a wildland firefighter with the US Forest Service in the spring and summer of 1977. More specifically, I had been hired as a seasonal helitack crewman (*helitack* refers to "helicopter-delivered fire resources").

At the time, Mike and I were living southeast of Albuquerque in the Manzanita (little apple) Mountains. The Manzanitas are a low-elevation bridge between the imposing Sandia (watermelon) mountains to the north and the Manzano (apple tree) mountains to the south. In 1976, we had naïvely purchased a charming but rustic cottage that was nestled among the pines and junipers on the level portion of five primarily sloping acres. The cottage's bay window looked out over a flower-filled meadow in summer and deep snows in winter. (Our elevation was about seven thousand feet.) Although the cottage did have running water, there was no well, and the water was supplied via a two thousand–gallon holding tank. A large truck wound its way up the twisting mountain road to bring us water in summer. But in winter, we hauled water in five-gallon jugs from the city because the truck couldn't always get through the snow, and, if not filled, the tank would freeze. We also melted snow for washing in the winter and huddled around our freestanding Franklin fireplace with our dogs. Needless to say, one winter of such rustic accommodations was enough, so we ended up selling our little house in the pines at the end of the fire season in 1977.

Had my firefighting position been based in the Sandia Ranger District of the Cibola National Forest, I would have had an easy commute. Alas, the position was instead based in the Mount Taylor Ranger District, in the faraway uranium town of Grants. This required me to drive down sinuous Highway 14 (actually heading north to the town of Tijeras), pick up Interstate 40 west through the incipiently sprawling city of Albuquerque, and head across the high and lonely mesas, eventually weaving through the *malpais* (lava beds) into Grants. The trip was one hundred miles each way and took about two hours without vehicle breakdowns.

Among our ever-changing stable of used cars was an old Fiat that had issues with its points. I would have to stop out in the middle of nowhere to reset those problematic points. Even our four-wheel drive Scout let me down on that trek across New Mexico, and I would have my pay docked because I was late to work. Thank God I was a mere twenty-six years old because I'm not at all sure that today I could make that two hundred–mile round-trip journey via unreliable conveyances on a nearly daily basis. Even after more than three decades as a field biologist, dubbing myself a "truck-driving mama" as I racked up the miles in Georgia and Florida, I realize that this 1977 commute was totally absurd.

A Marvelously Motley Crew

I've heard it said that soldiers in battlefield situations fight not only for their country and for freedom but also for their immediate comrades who share the battles. I have experienced this comradeship as a firefighter and later as a wildlife biologist struggling to conserve species in the face of development. The synergy created via my crew or team members was a valuable asset, especially when the going got rough.

My 1977 crew at the Sandpoint heliport in Grants was both diverse and colorful. Bruce was our fearless leader, and with his ample mustache and sideburns, he looked like a handsome Old West hero or gunslinger. Bruce taught me the usefulness of what he called "walk-talks," where he and the chosen person would stroll and discuss performance, problems, or perhaps just life. He was a solid crew boss, but his one real fault was occasional intense crabbiness that stemmed from the fasts he endured to maintain his weight. When his crankiness became intolerable, the crew would send me to talk to him about fasting while on a fire. For some reason, Bruce seemed to enjoy listening to me and often encouraged me to talk to him about anything when we would be stuck in the truck driving somewhere. Of course, I'm not sure how grateful he was for my interventions regarding his fasting, but he never reprimanded me for my candor.

Our assistant crew chief was John, a quiet, hardworking Native American. John was uber-reliable, and I had an enormous amount of respect for him. Other crew members included the only other female on our helitack crew, Susan, an earth mother who read us inspiring stories from *Tales of the Dervishes* and brought us healthy morsels to consume. When she first met fellow crew member Mark 1, who eerily resembled *Star Wars*' Han Solo/Harrison Ford in looks and attitude, she commented, "Oh, he's too good looking; I don't think I can work with him." As one of the few married folks on the crew, I replied, "Don't worry; you'll get over it." And she did, on the first fire when he exhibited his "it's all about me" MO, and we quickly learned we would have to crack the whip to get him to really work. Mark 2 was a bearded eco-dude (kindred spirit to me); Terry was the youngest, a hardworking college lad; and Dave was a gentle giant who was paired up with me on occasion because he was the heaviest and I was the lightest. Weight was always a concern when manifesting a helicopter (weight loading so that the chopper could safely get over the local eleven thousand–foot peaks).

Two other noteworthy characters who did not become part of our crew were Kathy, a beauty of Irish descent who struggled in her marriage to a traditional Pueblo Indian; and Albert, a fun-loving Hispanic who just had too much trouble with the math of manifesting the helicopter. Kathy became our primary lookout at La Mosca (the fly), high on Mount Taylor, and Albert found his place on a pumper (water truck) crew. I thought the world of all of them and wish that I had kept track of everyone after the fire season. Susan and I did write letters for a while, even after I headed east to grad school.

Intensive Training

Although not as elite or specialty trained as hotshots and smoke jumpers, helitack crew members are nevertheless a specialized, exceptionally well-trained subset of wildland firefighters. In a nutshell, helitack crews are dispatched to a fire via helicopter and serve as initial attack until pumper crews or other firefighters are called in to assist. On smaller fires within our own district, we often constituted the only firefighters, especially in remote, roadless areas where pumpers and ground crews could not easily gain access. Not only did we have to have the same firefighting acumen and training as other wildland firefighters; we had to know helicopters backward and forward. I was truly amazed at the depth of the training that seasonal helitack firefighters received. And for me, that training began in mid-April 1977, well before the height of fire season in the Southwest.

Understandably, fire behavior and suppression are the crux of what firefighters need to know. Wildland fires are strange, sometimes unpredictable, beasts that are subject to numerous variables of topography and weather. Second only to learning about fire behavior, suppression, and safety, I had to assimilate reams of information about helicopter types, manifesting, signaling (from the ground to assist the pilot in takeoffs and landings), and especially safety around these potentially dangerous conveyances. Our chopper was a French Lama, especially useful in high-altitude regions. It always looked somewhat like a large, white grasshopper to me. Larger Alouettes were also used, as were small, maneuverable Hughes 500s (what we called "flying sperms") and the ubiquitous Jet Rangers. Crew boss Bruce didn't like these slick choppers that you see in movies; he called them "Jet Dangers." I also learned about the workhorse whirlybird, the Sikorsky, and the big fixed-wing slurry bombers (B-52s, C-130s) that drop a pink-hued retardant on larger fires.

Our chopper training went well beyond mere identification or manifesting and loading. One of the scarier tasks involved so-called hover hookups. In this maneuver the helitack crewman has to quickly and efficiently scurry under a hovering chopper and attach a cable that suspends a slurry bucket to the helicopter's belly. Needless to say, we received extra hazard pay for this work. Once the bucket was successfully attached, the helicopter would head back toward the conflagration and release the retardant on hot spots that the slurry bombers might have missed or that were especially difficult to access. We were also expected to assist in refilling these buckets from hoses or from portable pools filled with the pink liquid. On one occasion, I received the universally recognized middle finger sign of disdain from Dorsey, our pilot, because I slipped and accidentally squirted the whirlybird windshield with slurry.

Besides the obvious instruction regarding everything you ever wanted to know (but were perhaps afraid to ask) about fire suppression and helicopters, we received training that would allow us to successfully employ portable radios, compasses, crosscut saws, chainsaws, and the three commonly used wildland firefighting tools: shovels, McLeods (hoe-like blade/tined blade combo), and Pulaskis (axe blade for chopping/pick-like blade for digging). And we needed to know how to sharpen and maintain these tools, including the chainsaws. Short courses in everything from ethics to first aid and defensive driving were also included.

But it was not only our beleaguered brains that received a workout: Our bodies had to be physically honed to a muscular and aerobic level of peak performance. Suffice it to say that, as a result, I reached my physical conditioning zenith at the tender age of twenty-six going on twenty-seven. I often likened the physical training to an Outward Bound course in some respects, though I have never taken one of those intensive courses. My first day's journal entry on April 11, 1977, noted at least ten types of exercises, including chin-ups (my undoing), and, of course, running. On most mornings, after driving for two hours, I ran a sandy course through the desert and arroyos (ditches); my only solace was the heartening song of the meadowlark. And I was not attired in running shorts, Nike shoes, and a wicks-sweat-away performance tee. Understandably, my workout was in firefighting clothes: cotton jeans, cotton long-sleeved shirt (no synthetics that might melt), and heavy work boots.

In April, I could run only a half mile in that excessively sandy, uneven terrain, but by early June, I ran the entire one and a half miles easily. We also had to negotiate an obstacle course (reminds me now of the CBS show *Survivor*, which of course didn't exist back then). I struggled with the eight-foot wall and broad jump. Because most women, especially thin ones like me, lack relative upper-body strength, I had to erect a chin-up bar outside my mountain cottage so that I could practice and achieve more than one and a half chin-ups per workout session and get over that huge wall.

Tackling Mount Taylor

Our ultimate physical test/training occurred in early May: the *big* hike up and over the 11,301-foot peak of Mount Taylor. Each of us carried sixty pounds (about half my body weight): forty pounds in a very uncomfortable Forest Service pack on our back and twenty pounds in our own pack in the front. The poorly padded backpack "just about killed my shoulders," as I noted in my journal.

We started at low elevation and hiked up a canyon along a stream. There was no real trail, and we bushwhacked uphill through brush. The brief lunch break of Kwik bars caused some roiling in my stomach, never a good sign when backpacking. I later wrote, "I barely made it up the last hill; felt like a sloth with sore shoulders." But the natural beauty of our eventual campsite in the tall pines overrode my physical discomfort, at least temporarily. Wild turkeys and mule deer made fleeting appearances, and a lone coyote yipped near camp. Despite my fatigue, sleep was as fleeting as the woodland creatures, and my paper sleeping bag did little to protect me from a temperature of 18°F. Around 2:30 a.m., we arose and built (rather than suppressed) a small fire to keep warm.

The second day of the marathon hike dawned beautifully but portended to be more challenging. At daybreak, I looked up to see cattle surrounding the camp as if we were cowboys in some old western movie. Breakfast consisted of C rations (military ready-to-eat meals); I had noted that the Forest Service rations (commercial brands of spaghetti) from the night before were tastier. Our morning's training was on the use of two-man crosscut saws and chainsaws. Let's just say that there is a reason some wise man thought to create the chainsaw. The crosscut saws that the old lumbermen used require considerable upper-body strength, stamina, and obvious coordination between sawyers. Regarding my initial attempt at using a chainsaw, I wrote: "I screwed up but did OK considering it was my first time." By 2:00 p.m., I was bushed—but we still had a mountain to conquer. Our pack weight had been reduced to forty-five pounds, but the going was still rough at the considerably higher altitude. Painstakingly, we made our way up Mount Taylor. My journal reflected the strenuous ascent: "I went *very slow*, slowest in fact."

John, our assistant crew boss, kindly waited with me as I literally went into slow-motion mode above eleven thousand feet. I think I told him to use a crop on me so that I would go faster, like a recalcitrant horse. But my legs and lungs determined my pace, and eventually we did top out to a stunning view and strong wind. The steep descent on the backside of the mountain provided little relief as we stumbled through snowdrifts (northern exposure) and finally made it back to the truck ("a dear sight") before 6:00 p.m. I had summed up the hike's conclusion succinctly: "I'm one big bruise but I made it!"

Practice Makes Perfect . . . or Not

In addition to classroom instruction, hands-on equipment training, and extreme physical conditioning, we were given our own practice fires to extinguish. What an eye-opener that was. Albert and I screwed up with style. He spent too much time in one area (which allowed the fire to escape in other areas), and I tried to dig a fire line in the dirt instead of knocking the flames down first. Consequently, both fires took off and the entire team had to bust our butts to stop them. Exhausting doesn't even begin to describe this early-morning ego deflator. We couldn't wait for lunchtime, and when it finally arrived, we all just flopped down under a tree and were almost too tired to eat.

In the afternoon, we learned the fine art of mopping up a fire. When wildland firefighters walk, drive, or fly away from a fire scene, there can be absolutely no smokes, leftover hot spots, smoldering snags, or any other indication that a dreaded reburn might occur. On smaller fires, that means actually testing the burned earth with your bare hands. The fear of God (and the government) was drummed into our minds and seared into our souls: *You never want to be responsible for a reburn.*

Our crew chief, Bruce, recounted the tale of members of a previous helitack crew who had walked away from a fire before it was thoroughly mopped up and was truly out: Their lives and careers were shattered by a lasting stigma. We were even taken to the site of this reburn on Mount Taylor during our over-the-mountain hike. The scars on the land and apparently on the errant firefighters remained for many years.

First Real Fires

So, after the training and practice, we were ready, right? We were badass firefighters who could save the forests and mountain meadows. My first actual fire occurred in mid-May. The fire call came in just before lunchtime, and Bruce, Mark 2, and I scrambled to board the "ship" and flew west to the Zuni Mountains. This was a small fire, about an acre, that had been started by a lightning strike. It was a creeper that should have been relatively easy to suppress. Alas, that was not the case.

First, we stomped the fire's edge and then dug a fire line to mineral soil with our Pulaski tools. The rocky terrain made digging an arduous undertaking. Then, Mark and I chopped small oak trees with the axe side of our Pulaski until we could no longer stand, so we continued chopping on our knees. I had not eaten since breakfast, and I have always lived close to my basal metabolism, meaning *I need food*. Truly, I had never been so tired in my life. We kept up the intense action for more than six hours with no break. Finally, a water tanker truck and crew laboriously made their way to the remote site so that we helitack folks could fly back to our base. Somewhat embarrassingly, I was so stiff the next day that I could barely move.

Three days after my first fire, a call came in late in the afternoon.

Terry and I scurried to the chopper, which was already loaded with our equipment and gear. Our flight to the northern side of Mount Taylor revealed stunningly beautiful scenery. The fire was at the northern edge of our district and was a blessedly small, smoldering fire that was somewhat amoeba shaped, with long fingers reaching into unburned areas. There was a local witness present, but he spoke only Spanish and we spoke only English. (This was the first of many times that I wished I had taken Spanish instead of French in high school.)

Terry and I started a fire line at the top of a hill and dug for three hours until a tanker crew showed up. The five of us now dug until 11:30 that night, ate a quick C ration, and worked on hot spots with portable backpack pumps ("piss pumps") until 2:30 a.m. I noted that it was weird and a bit eerie working at night with headlamps, although the amazing canopy of stars provided solace. The night was freezing cold, and when we finally slipped into our paper sleeping bags, sleep came only because we were totally spent.

The next morning proved to be really cold, with a typical New Mexico spring wind increasing in speed, and we mopped up (assuring that all little smokes were suppressed) until early afternoon, with only a quick C-ration break. One of the tanker crew members left to look for a compatriot in a second truck, who had gotten lost coming to retrieve us. We waited two more hours to be sure the fire was truly out and then started the long drive back to base.

It was not atypical for us to have to find our own way back to the heliport. Depending on the remoteness of the location, the chopper didn't always come to retrieve us. The shadows were long and it was after 6:00 p.m. when I started my long commute home. To stay awake, I ate an apple very slowly and then sang at the top of my voice; back then, radio stations were few and far between along that lonely stretch of I-40.

Things Heat Up

Memorial Day 1977 dawned bright and breezeless, with the promise of searing heat later in the day. I was sore from my attempts at being a cowgirl the day before. A neighbor near my mountain cottage had just acquired a real cowpony and wanted some help gathering his wandering cows. I jumped at the chance but quickly realized that this rodeo horse had moves like Mick Jagger. It was all I could do to remain in the saddle, as the horse would zero in on an errant cow and then feint right or left with the bovine's movements. When I got to the heliport on that Monday morning, I could barely hobble around one lap of the sandy arroyos that constituted our running trail. After lunch, a telltale smoke arose over Mount Taylor, and Bruce, Mark 2, and I headed to a human-caused fire in Lobo Canyon that had moved onto private land.

I recall the staggering heat from both the sun and the fire as I exited the chopper. We wore yellow jumpsuits over our fire shirts and jeans. It was sweltering, even after I tore off the jumpsuit, and I hoped that I wouldn't embarrass myself by passing out. The terrain was incredibly rugged; we dug line literally among the rocks. One of my most vivid and touching memories occurred when I looked up to find an older man, still in his white office shirt with the sleeves rolled up, digging beside me on the line. It was his land, and he rushed to help us as soon as he got home from work. The local landowners were extremely grateful that we were there and showed their gratitude by bringing us delicious chicken and fruit that evening. Real food, rather than C rations or Forest Service rations, was a godsend.

We dug line until 9:00 p.m. and then turned the fire over to a pumper crew to mop up. The heat had subsided. It was a stunning moonlit night as we made our way back to the heliport in one of the crew trucks. It was much too late to drive home, so I rolled out a sleeping bag at our aptly named Sandpoint and slept with the mice. It was a Memorial Day to remember, for we had fought a battle to save folks' land and homes—and we had prevailed.

The fire season was heating up, so to speak, and on the first day of June, our crew split up to fight two different fires. Bruce, Susan, and I headed out to a fire on state land. I chopped trees that arched over the fire line, and we used fire to fight fire by burning out swaths of vegetation. However, my primary job was to scout the fire line for any flames that might have made the jump to the other side. The state folks had called for slurry bombers to help contain this fire, and I got an admittedly scary chance to use my earlier training. We had been instructed to seriously avoid getting under the dump of fire retardant from a bomber. But in this particular case, I was scouting far down the fire line away from my crew when the bomber seemed to swoop down out of nowhere and released its load just upwind from me. I quickly went into the fetal position on the ground, as we had been trained to do, and thankfully felt only drops of retardant as the trees and earth turned a surreal pink from the drop. There was no way to deny having been "slurried": I had worn off-white jeans that day, and they now had pink blotches that never did wash out.

The slides that we had been shown during our pre-fire-season training showed trucks being damaged by being in the direct line of a drop. I had been quite lucky that day; the bomber was higher than usual due to tall trees, and I had dodged the worst of the pink "rain." Because Susan had been digging line for hours, we put her to bed at 9:00 p.m., and Bruce and I dug until 1:00 a.m. He was unusually tired because he had been fasting for several days (at some point, I did have that talk with my boss about not fasting when engaged in active firefighting). Once again, our willingness to assist others paid off: The state fire crew brought us sandwiches and other delightfully yummy goodies. We turned the fire over to them and slept the sleep of the genuinely exhausted under a moonlit sky.

The next morning, we walked the line to assure there was no fire encroachment over it and then were forced to hike out to an open area where the helicopter could land. Apparently, our pumper crews were deployed elsewhere, and the remote location warranted the chopper returning to retrieve us. We were each carrying forty to fifty pounds of gear, which is the reason that our original training had us hiking with such heavy packs. Bruce climbed a tree in an attempt to find open ground, but we still had to wait until we heard the chopper before we could find that meadow.

When we got back to base, I had time for only a quick shower to get the ash and slurry off me before the crew was divided up and dispatched to two different smokes on the eastern side of Mount Taylor. Thunderstorms were cracking, and the rain beat down on the windshield of the chopper as we flew to the assigned fire. My journal succinctly described the fire: "Ours was a bitch!" The fire was burning on an incredibly steep hill, and the initial attack was absolutely grueling. With the unexpected but welcomed assistance of a local rancher and Bureau of Land Management (BLM) employees, we got the line dug. I luckily was able to work the top of the hill for a while. At 11:00 p.m., a pumper crew arrived and we broke for some much-needed nourishment. We slept in shifts; mine was 12:00 to 4:00 a.m. When I resumed the fight, I was working on the slope and fell down the hill but thankfully was not injured. This was a *wicked witch-bitch* fire, for sure. We mopped up all morning and into the afternoon, and I had noted: "This was my roughest fire yet; that hill was driving me crazy!"

Finally, some blessed rain helped with our mop-up chores. We worked until 7:00 p.m., were relieved by a pumper crew, and had a two-hour drive back to base, soaked to the skin. After a shower and some pizza, I barely made it home at 3:30 a.m. My journal entry was brief: *"Tired!"* In retrospect, I truly don't know how I fought fires under those intense conditions and then drove all the way back to Albuquerque and beyond. This stint as a firefighter was truly the ideal test and training for the making of a wildlife biologist.

If the steep hill fire (called the Tapia Fire, because all wildland fires are named) was my roughest fire, the Lost Fire proved to be potentially the most dangerous. After sleeping in the mountains above the heliport, I awoke to the gentle patter of light rain on my contrary sleeping bag that had refused to zip throughout the night. Truly, sleeping bags that won't zip are among the most frustrating issues associated with camping. But the rain provided a soothing meditation and gave no hint of the drama to come. I at least was able to get breakfast at a local café before the call came in for a fire in the Zuni Mountains.

Assistant Crew Chief John, Mark 1, and I were dropped off on an initially small but rapidly escalating fire. We vigorously attacked it, but the squirrelly canyon winds caused it to blow up so quickly that it elicited both awe and extreme terror. The fire torched out and morphed into an uncontrollable beast in the crowns of the towering pines. Our only recourse was retreat. This was a monster that we couldn't vanquish with our small force. I remember looking back as we ran and saw that the fire had reached our chainsaws, which exploded like fireworks on the Fourth of July. Fortunately, we received reinforcements via a pumper crew and local Navajos but not before the crazed fire jumped an arroyo and consumed the rest of our gear. I lost my beloved down jacket, but the five dollars stashed in the pocket somehow survived in the melted coat (as poor as I was then, I was pleased that at least I had five dollars).

With sundown and the reinforcements, we calmed the flaming beast, slept a few hours, and then mopped up with piss pumps in the wee hours. We watched the sun rise as we worked on the hot spots, and eventually we were released to make our way back to the heliport with the pumper crew. As usual, I was teased about being so dirty. Apparently, I have a tendency to wipe my sweaty face with my sooty (in this case) or sandy (in the case of my wildlife work) hands and end up with just two blue eyes staring out of a filthy face. One of my wildlife colleagues later dubbed me "Joan Dirt" after that goofy David Spade movie *Joe Dirt* graced (or is it disgraced?) the movie theaters.

Arizona on Fire

It was only mid-June and I had survived fires that generated mind- and body-numbing fatigue and intense terror. If my roughest and scariest fires were behind me, my longest deployment lay ahead. After a hot and muggy ("muggy" is a rare descriptive term in the Southwest) day of mundane tasks at the heliport, I ferried Mark 1 back to Albuquerque for a rendezvous with his girlfriend. I arrived at my mountain cottage between 9:30 and 10:00 p.m., only to receive a call from John saying that we'd been dispatched to Arizona. What? I now had to find Mark before he went out dancing or whatever, leave a note for Mike, and quickly change clothes and pack for who knew how long. I located Mark and sped back to Grants, whereupon John, Susan, and I loaded up the truck and headed for Arizona after midnight. This of course was way before cell phones, so I had to call a not-too-thrilled Mike when we stopped in Flagstaff.

We continued up to the North Rim of the Grand Canyon, taking shifts driving and catching minimal shut-eye. Shortly after we arrived at the staging area that overlooked a vast valley with flaming forests, we saw our chopper descend with Bruce and pilot Dorsey. We quickly set up for bucketloads of slurry and learned how to run in and fill the bucket with a hose rather than dipping it into a pool. On off-district fires, we were not initial attack but instead provided helitack support via slurry drops on hot spots that the larger bombers missed. From our hilltop vantage point, we could see the war being waged by the bombers as they dumped megaloads of pink slurry on the inferno. At one point, the fire "spotted" nearly a mile across a valley. This fire was too dangerous to send in ground troops yet; it was up to the aerial forces to subdue it.

The valleys and vast ponderosa pine forests north of the "Big Ditch," the Grand Canyon, are magnificent, even on fire, and I noted the beauty in my journal. I learned a great deal that day, for this was my first off-district fire. But it wasn't going to be my last because fire season was now in full swing.

From the North Rim, we were quickly dispatched to yet another tree-gobbling Arizona fire, and I was the lucky one who got to fly with Bruce and Dorsey over the Grand Canyon at sunset (truly awe inspiring). Poor John and Susan had to drive all the way back to Flagstaff and had a flat tire, so they didn't arrive until 2:00 a.m. None of us got much sleep before getting up in the wee hours to head south to the Mogollon Rim country and its vast ponderosa pine forests. Again, I lucked out and got to fly, noting the thick smoke as we descended into the designated heliport.

This was a large fire camp. There were Arizona crews from Prescott and Payson in addition to our New Mexico–based crew. These camps are like temporary, mega-efficient, military outposts, and we got a good hot breakfast (not C rations) in this beautiful location. Our crew worked slurry buckets again, and we got to tell tales and compare notes with other helitack crews later in the evening.

I think this was the fire where I was caught off guard by bigotry against Native Americans. Most of the helitack men and women I met were fine folks; that said, jerks exist in every profession. As we chatted in the dark that night, inane and insensitive comments about "Indians" came from one or two idiots. John was of Native descent, and I was sorry that he had to listen to this detritus. I seem to recall that one of our crew made a comment about John being an Indian, and that seemed to quickly quell the brainless remarks. Heretofore, I had not really seen prejudice on fires, either based on ethnicity or gender. The Southwest is a human *masala*, a fascinating mix of Native Americans, Hispanics, and Anglos. The cultures are rich and woven throughout the region's long history; unfortunately, the so-called winning of the West is tarnished by too many tales of bigotry. Stupidity, complacency, and prejudice are extremely distasteful and difficult traits to extinguish in any era.

Our initial off-district adventure was only beginning. My journal reflected an appreciation for the little things during these often demanding deployments: "Had fantastic breakfast at heliport (chile cheese omelets, etc.)." After breakfast, Susan got the chopper ride while John and I drove to Flagstaff for a week-long battle to vanquish the giant Radio Fire that was threatening the city. Mount Elden, which looms above Flagstaff, was ablaze: Four thousand acres were burning; the previous two fires were six hundred to a thousand acres. Our heliport and home base was Buffalo Park, a large, rocky meadow that sat at the foot of Mount Elden. We worked in concert with five other choppers and crews, two of which were the elite helitack crews from California. We tried not to have equipment envy, but those California crews were outfitted to the hilt.

The days meshed together as we would arise before dawn and fill slurry buckets for hot spots. We were lucky to be able to eat in town and stay in motels. At one point, when I took the crew's clothes to a local Laundromat, the manager let me wash up there and conveyed sincere thanks to all the firefighters for saving the city.

The Radio Fire turned out to be big news not only in Flagstaff but throughout Arizona, and we had our hands full trying to get it under control. The late June weather was not helping; on at least one day, we were shut down by extreme winds. At that time, we couldn't fly if the winds were greater than thirty miles per hour. Only the largest of the California helicopters continued the fight along with the bombers. I remember some of our crew lying like tuckered pups in the truck, with little regard as to whose leg or arm rested on another's. By that point, we were family, and many of the privacy boundaries that define daily life in the real world tend to dissipate in these extreme conditions. That said, Susan and I did have to figure out how to pee in this vast, treeless meadow. Finally, we simply opened the door of the truck and squatted behind it.

Murphy's Law often prevails on these large fires, and on June 20, everything really did go wrong. The Alouette chopper accidentally crushed its slurry bucket, and the cables attaching the buckets to our Lama and the Hughes chopper snapped. I felt terrible about our cable because I was signaling and didn't see John's "hold" sign, which translated to "Don't lift the damn bucket yet!" We spent most of the morning fixing it, and I was truly bummed that I had screwed up.

To add to the equipment malfunctions, both Bruce and I had upset stomachs. Fortunately, things improved later in the week, though I did have to go into town to wire Mike for money. This deployment was dragging on, for sure. At least the evenings were sometimes interesting. We would go to a honky-tonk named Granny's Closet, where we would dance until after midnight if we weren't too tired. (Remember that I was only twenty-six then, and most of the crew were even younger.)

For many reasons, I had hoped to get home by June 24, my fifth wedding anniversary. Wildland firefighting can be hard on marriages, and comments in my journal like, "Caught hell as usual" and "Mike was ticked off for two days: What a life!" conveyed that stress. Mike had wanted a conventional wife—and he got the antithesis. For some reason, he thought I would change and that the adventurous spirit that had initially attracted him would abate. That spirit only burned brighter as time went on. So I felt it was really important to make it home on my anniversary and was heartened when I learned we were being released.

Alas, we didn't get far: We had to spend all afternoon at the Flagstaff airport ferrying hotshot firefighters on a rainy day where lightning fires were popping up. Due to a snafu with another chopper that knocked it out of commission, we did not get to leave at 6:00 p.m. but instead had to spend another night. I had to call my husband back and tell him that our anniversary celebration was a no-go. My crew and other new comrades felt bad for me, so Bruce rented a car and twelve of us went to the Cowboy Country Club. I must confess to having a ball; we dined on steak, beans, and Indian fry bread, and washed it down with strawberry daiquiris. And we danced: the "shuffle, rock, twist, slow—*fun!*" I was impressed by the western decor, complete with saddles, and I was touched when my crew gave me earrings.

We finally did escape early the next morning, and even though I had known the California crew members only a short time, we had bonded to the point where good-byes were sad. I got to fly back on the chopper and enjoyed amazing views of Meteor Crater, Painted Desert, and Petrified Forest in eastern Arizona. However, the ride became a whole lot less fun in New Mexico, as we hit turbulent air and I felt a twinge of airsickness. My anticipation of returning to our own heliport was definitely dampened when we were immediately dispatched to a fire on the edge of the Mount Taylor District. The helicopter stayed with us this time, and Bruce, Terry, and I attacked the small but tenacious fire. We had to really work hard to finish up before it was too dark to fly back. (Some choppers are equipped to fly at night, but ours was not.) Of course, I still had a two-hour drive ahead of me, after showering to remove the pervasive ash and sweat. Some ten days after I had rushed out of my mountain abode to head to the Grand Canyon State, I was finally home.

Curious Happenstances

Of course, there was little time to rest and recuperate. Several days after returning from our big Arizona adventure, we were dispatched in the wee hours to assist with a fire near the tiny New Mexico town of Magdalena, southwest of Albuquerque. A warm, moonlit night gave way to a blazingly hot day. By some accounts, the firefighting need had been overestimated, with both slurry drops and hotshot crews called in on an approximately forty-five-acre fire. We spent the day ferrying crews. Interestingly, as well trained as hotshots and smokejumpers are, we always reminded them to keep their heads down when approaching or exiting the helicopters. Crew Chief Bruce was adamant that we grab these often hulking guys by the back of their necks and hold their heads down until they were beyond the reach of the blades. I always explained what I was doing and why I was doing it, because grabbing someone's neck is a bit personal, to say the least.

As if neck grabbing weren't unusual enough, I had noted in my journal that this was a "weird day": we found a Mexican black-tailed rattlesnake on the exact point where the choppers were to land. Mark 2 and I provided the "eco-freak" opinion that we should shoo the snake away and let it live. However, the prevailing opinion was that the guys had to kill it to protect people who wouldn't be able to hear the rattle with the deafening sound of the chopper. The demise of this magnificent snake prompted a spirited discussion about eco-freaks and conservation on a rainbow-enhanced drive back to the small and quaint town of Magdalena. The killing of the snake was only one of the unusual incidents related to this two-day deployment: A girl fell down the thunderstorm-slick steps of a lookout and had to be taken to the hospital, and a fire prevention technician rolled his truck and also ended up in the hospital. Thankfully, neither sustained serious injuries.

On the drive back to Grants, we also stopped to help some folks with a malfunctioning vehicle in the middle of nowhere, but they graciously declined our assistance because a tow truck was already on its way. Exhausted, I curled up on the front seat of the truck as we made our way back to our heliport. Because it was after midnight when we pulled into Sandpoint, I slept—or tried to sleep—in the upper parking lot. My much-needed slumber was interrupted by mosquitoes (yes, we can get those pesky bloodsuckers in the Southwest) and a roaming skunk. I finally had to retreat to my old Fiat until 5:00 a.m., and because I didn't have a key to the building, I was forced to crawl through the window to get to a phone and call Mike. For many reasons (including an after-work crew party), I didn't get home until 2:30 a.m. the following morning. I didn't feel quite so guilty because my hubby had plans in the city that evening anyway. Obviously though, Mike and I were on different wavelengths that could often seem like different planets.

Firefighting, like life, is full of ups and downs; unpredictability is part of the equation. On a hot and miserable day in July, the crew was unanimously irked at bossman Bruce, who decided we needed to do a "burnout" physical fitness challenge. Everyone was in a rotten mood; I was "disgusted." I ran a mile and three-quarters, did sixty-two sit-ups (apparently not good enough), and struggled with twenty-two push-ups. And then, as we were licking our wounds, so to speak, a fire call came in late in the day.

The three-acre fire was creeping through a stunningly beautiful area of the Zuni Mountains. Mark 1 and I did initial attack, backed up by the rest of the helitack crew and three pumpers. We dug line until 11:30 p.m., ate dinner, and then bedded down in shifts near quietly burning logs on a typically cool mountain night. My journal reflected this more gentlemanly effort: "*Neat* fire!" In the morning, we ate our C rations and mopped up until noon. I dug out one smoldering stump for an hour—but still appreciated this slower-pace mop-up. The drive back to Grants wound through the gleaming white trunks and shimmering leaves of aspen groves. Thankfully, not all fire fights were nightmarish drudgery.

Change Is Afoot

As the summer wore on and the monsoon season provided much-needed moisture, firefighting needs on our Mount Taylor District were reduced. The helicopter was dispatched elsewhere, and some crew members were moving on to their next job or schooling. Those of us who remained spent long days "stacking sticks"—reducing fuel loads in strategic areas of the forest. Apparently, the months of inhaling smoke had taken its toll, and I was suffering from allergies that made me as sleepy as the *Wizard of Oz*'s Dorothy in the poppy patch. I also was having some heart palpitations, which were eventually diagnosed as premature ventricular contractions (PVCs). Although not uncommon, PVCs are still disconcerting. I was told that they can occur in young folks that are in otherwise great shape. But the combination of sleep-inducing allergies, PVCs, and a husband who was rapidly losing patience with my unconventional job led me to hang up my firefighting tools and say adios to an amazing and unforgettable fire season. My remaining crew members were dispatched to megafires in California, and I felt incredibly left behind. This surely wasn't the way I had wanted the summer to end, but it was time to move on to the next chapter of *mi vida loca* (my crazy life).

Autumn 1977 brought new challenges. Mike and I had sold our little cottage in the mountains (one winter of melting snow and hauling gallons of water from the city was enough), and I came to the realization that I would need to go back to school to achieve a career in wildlife biology. Thus, another quest was born: We set out on a cross-country trip to look at graduate schools for me and to find a locale and institution where Mike could complete his undergrad degree. I quickly got an education about sexual discrimination in academia. At a university in Michigan, a senior professor briefly listened to my interests and aspirations and had the nerve to sarcastically comment, "Wouldn't you just be happier back at your stove?" Even Mike, hardly a proponent of women's liberation, was floored by that snarky inquiry.

At a Utah college, the wildlife professor who might have been able to sponsor me had just been mauled by a grizzly bear. Mike was sitting in the waiting area when this horribly disfigured man walked by. He was in no condition at the moment to take on new grad students. At other universities in the Rocky Mountain West, the reality that wildlife biology was indeed a male bastion was grilled into my naïve psyche. I returned to a rental house in Albuquerque with my tail figuratively tucked between my legs. But after all I had been through, I wasn't about to give up. Tenacity would continue to be my strong suit.

2

Metamorphosis: From Snake Lady to Gopher Queen

Press on.... Persistence and determination alone are omnipotent.

—CALVIN COOLIDGE

7

Indigo Daze

The Snake Lady Goes Down to Georgia

The Search Yields Fruit

Despite having sexist doors slammed in my face, my quest to find a graduate wildlife program was hardly over. I contacted my undergraduate wildlife professor in Kentucky, and he suggested I try three schools in the Southeast with excellent wildlife programs: Auburn University (Alabama); VPI (now Virginia Tech); and LSU (Louisiana State University) down in Cajun country. I immediately sent letters of inquiry to all three universities. An encouraging and intriguing letter came back from the Alabama Cooperative Wildlife Research Unit at Auburn. My academic record, experience, and the recommendation of my former wildlife professor apparently piqued the curiosity of the head of the Research Unit, a co-op of federal, state, university, and Wildlife Management Institute funding. Would I be interested in a study of eastern indigo snake distribution in southern Georgia?

A grant was forthcoming from the Georgia Department of Natural Resources, and the study would involve questionnaires, interviews, soil and habitat mapping, fieldwork, and a "good bit of travel" throughout the Georgia Coastal Plain. Zoogeography, the science of geographic distribution of animal species, had always been an interest of mine. But I had never even heard of the indigo snake, and of course, Google didn't exist back then, so I scrambled to learn everything I could about this threatened species.

Fortunately for me, I had never been ophidiophobic (fearful of snakes). On the contrary, I was fascinated with these unfairly maligned and sinuous so-called symbols of original sin. They are beautiful, mysterious, and useful, in that they consume many of the pests that infiltrate our crops and homes. And I quickly learned that the indigo is in a top snaky echelon: It is nonvenomous, docile, reclusive, and spectacular in color and size; and it consumes rodents and venomous snakes. Even those who don't have any use for other species of snakes can often be convinced to protect indigos. My early sleuthing regarding the science of the eastern indigo, *Drymarchon corais couperi* (now known as just *D. couperi*), revealed that it is one of the largest nonvenomous snakes in North America and reaches lengths of eight and a half feet.

As its common name bespeaks, it is a stunning blue-black hue, with smooth scales that reflect an iridescent sheen in the sunlight. Some individuals have a brilliant coral, rust, or cream suffusion on their throats and chins. I also learned that despite these seemingly distinctive attributes, the eastern indigo snake is often mistaken for other species within its range. Smaller specimens especially may resemble the common black racer, a slender, fast-moving nonvenomous snake with a white chin and a less glossy coloration. The indigo snake's other common names are gopher snake and blue gopher, because it inhabits the burrows of the gopher tortoise.

Restricted to the Coastal Plain of the southeastern United States, the eastern indigo was once reported to occur from South Carolina to Florida and westward to southern Louisiana. However, germane to my impending proposed study, the current range was considered to be Florida and southeastern Georgia, with little detail on either its precise distribution or habitat in Georgia. And while it apparently uses a variety of dry and wet habitats in warm, balmy Florida, its habitat in southern Georgia was more restrictive, being originally described as the "dry pine hills," also known as sandhills. Obviously, there were many gaps in the knowledge about indigo snakes, and the concern that it was declining had heightened the interest in this secretive species that often dwelled deep inside the earth.

Habitat destruction associated with real estate development, certain forestry practices, and agriculture was adversely affecting both the indigo snake and its benefactor, the gopher tortoise. I also learned that the practice of introducing gasoline into tortoise burrows to drive out rattlesnakes (that were then taken to roundups) was likely fatal to indigos. And ironically, those who admired this reptile contributed to its decline through overcollection for pets (yes, *pets*), while those who abhorred it deliberately killed this beneficial species because of the misguided belief that "the only good snake is a dead snake."

Armed with some semblance of knowledge about eastern indigo snakes, I wholeheartedly pushed forward to gain acceptance at Auburn University and to undertake this intriguing study. Winter 1977 to 1978 was a busy blur as I provided Auburn with all the necessary academic information. And by February 1978, I was finally accepted into a graduate program in wildlife biology. So did I choose Auburn University, or did it choose me?

I will always believe that it was kismet, fate, that allowed me to pursue my graduate work at this august institution and to be a part of a study that combined biology, geography, sociology, and a pinch of psychology. There is a branch of science called ethnozoology, the study of past and present interrelationships between human cultures and the animals in their environment. The indigo snake study set the stage for me to become immersed in this fascinating science throughout my career. I was about to learn a valuable lesson about wildlife biologists: Although we may think we want to be "woods hermits," wandering the forests and fields in pursuit of knowledge or providing assistance to wild populations, the truth is that we manage a triad of wildlife, habitats, and humans.

Third Transition: Alabama Bound

I honestly don't recall exactly what conventional hubby Mike initially thought about this atypical undertaking. Later, it became quite obvious that he was less than enthralled with a wife who was roaming southern Georgia in pursuit of a snake. Because he was employed in sales at a prominent brick company in Albuquerque (having finished for the time being with the military), he opted to stay behind initially until I could see how this adventure was going to unfold. But the plan was that he would eventually join me and finish his undergraduate degree.

So one March day, I alone pulled out of my driveway, temporarily leaving behind a likely somewhat confused husband and two much-loved canines, and began a lengthy cross-country trip to my new life. At the time, I really had no concept of how Auburn, Alabama, would look. I knew Texas and Oklahoma but had little knowledge of the Deep South. What I do vividly remember is passing through Montgomery, Alabama's, Black Belt and worrying that Auburn would be nothing but vast, seemingly endless, cotton fields. I was incredibly relieved when I began to see tall pines (yes, *trees*) as I was coming into the charming village of Auburn.

Initially, I rented a tiny, one-bedroom apartment just north of the university. But because I anticipated the eventual arrival of my husband, I wanted to find a house with more space and a yard. Money was of course an issue, and I wasn't turning up much that I could afford in Auburn itself. And then I stumbled upon a pleasant, moderate-sized, two-bedroom house with a fenced yard in the adjoining mill town of Opelika. The landlady was a jewel who was willing to work with a poor grad student. And although I did have to drive farther, the lower rent and the quasi-rural setting sealed the deal. As an unexpected perk, the movie *Norma Rae* (which later became one of my all-time favorites and which snagged Sally Field her first, much-deserved, Oscar) was being filmed while I lived in Opelika. Some of the cast and crew were staying at the Golden Cherry motel in downtown Opelika, one of the many filming locales in the area.

Unusual Neighbors

Living in Opelika was about to give me a profound cultural education. My new next-door neighbors worked at the local textile mill, and Sarah tried out as an extra in the movie, but she fainted from the oppressive heat inside one of the mill buildings and had to be taken out. Her kind and even-keeled husband, Lawrence, was an extremely hard worker and became an excellent neighbor who fed my two canine kids when I had to travel for my research. (Yes, I did go back to New Mexico to retrieve my dogs, Benji and Elsa.) Sarah was more mercurial and easily agitated, but she was also a gracious neighbor who taught me to eat a smorgasbord of Southern cuisine. One of her specialties was poke salad (pokeweed, *Phytolacca americana*), which can be dangerous if not cooked properly. I wouldn't have trusted anyone but Sarah to fix it for me. Both Sarah and Lawrence of course thought I was too skinny, so I received many fine Southern meats and veggies. Unfortunately, wayward stepchildren complicated my neighbors' life and created sometimes strange domestic friction. In an unsettling case early one Sunday morning, my phone was used to call the police when Sarah's adult son apparently began chasing Lawrence around the house with a hammer.

But the neighborhood eccentricities extended well beyond my next-door neighbors. Early in my Opelika experience, I would hear bizarre hollering that seemed to be moving up and down the road in front of my house. As a busy grad student, I wasn't home much; but at times, I would catch a glimpse of someone pulling a wagon, followed by some sort of hound dog. Inquiries to the neighbors revealed that I was hearing "ole Pete," a "screamer." Pete was apparently a forty-something-year-old, developmentally disabled man who lived with his mother. Pete would occasionally "escape" with his wagon and dog to walk the local back roads and, well, scream.

The sound was unsettling, to say the least, but knowing the source and a bit of the history made it easier to consider this odd ritual a part of the local color. If I saw Pete as I was driving in or out, I would wave. But I had never had any true contact with him until one fateful weekday morning as I was rushing around to get to class on time. To say I was surprised to find Pete standing at my door is an understatement. He mumbled something about needing to use my phone, and against my better judgment, I let him in. I think he did use the phone (I was making my lunch or was otherwise engaged), but then I couldn't find him. He had wandered into my guest bedroom, where I kept my outerwear, and was fondling my sweaters. I can only imagine the look on my face—and I think he said something like "I *like* sweaters." It took a few moments, but I eventually got him out of the house, and I rushed off to class. I was definitely unnerved, and when I later recounted the story to Lawrence, he laughed heartily and said, "Didn't anyone ever tell you that Pete has a sweater fetish? *Never* let him in the house, and don't *ever* leave sweaters out on the clothesline."

Inimitable Dr. Dan

I would soon realize that both Alabama and Georgia have more than their share of colorful characters. And one of the most colorful was my major professor and the leader of Auburn's Wildlife Research Unit. Alabama born and raised, Dr. Dan was indeed a legend in his own time. Stories abounded regarding his wild, or should I say wilder, days. He was in his early fifties by the time I met him, but his mind was still "down in the primordial slime," as one of my field buddies, Joe, use to say. For example, nearly every time we drove by a Badcock furniture store, he would exclaim (in his thick Southern accent), "Why would anyone name a store Baaaad Cock?" Southern manners and decorum not only didn't exist; they were totally out the window with Dr. Dan. Much of the time, he had a fun-loving nature and a wicked sense of humor. When afield, he repeatedly—and gleefully—pointed out the plant known as queen's delight (which was undeniably quite phallic in appearance) and couldn't resist talking about "those horny botanists," especially when we were looking at the vine *Clitoria*. (Georgia O'Keeffe would have loved painting this flower.) And when we headed back to civilization after a long day in the field, he would sometimes ask if we needed anyone to check us for ticks. (Many years later, a similar phrase turned up in a tongue-in-cheek Brad Paisley country song.)

If Dr. Dan overheard the grad students talking about how to afford going to scientific meetings, he would mischievously comment, "You can all get one *big* bed." Let me say at the outset that he never came on to me or any other student or tech; he simply embraced sexuality as another comical part of life. Moreover, he was happily married to an amazing woman, Ruth (or Baby Ruth, as he called her). Ruth was a stellar botanist, herbalist (a *curandera* of sorts), and animal lover. She became my staunch supporter, friend, and ally. She of course had her hands full dealing with Dr. Dan. One day when she and I returned from a shopping foray and found blueberry pie smeared all over the kitchen floor and counter, she asked Dr. Dan what had happened. He cussed and carried on that he spilled the pie but had cleaned it up. She turned to me and commented that even after paying for him to get a PhD, he still had a four-letter-word vocabulary.

I was one of Dr. Dan's first female grad students; in fact, the Wildlife Research Unit apparently had had only one other woman grad student previously. She was supposed to work on owls but was allegedly fearful of going out alone after dark. I don't know if Dr. Dan worried about my safety, but his unusual mentoring style included many cautionary tales. For example, he told me stories of rough rednecks and a corrupt sheriff who allegedly kidnapped young women for the sex-slave trade in the mysterious tiny town of Ludowici, Georgia, which was within my ninety-four-county study area. Actually, there was some truth to that part of Georgia being a hotbed of illegal activity and strange happenings during the 1950s and 1960s. Regardless of the veracity of the stories, I was always hyperalert when driving through Ludowici and working in the surrounding area.

Dr. Dan also cautioned me about the vastness of the Georgia piney woods, where one could easily become lost in a maze of thick palmetto and hurrah bush, so named because you had to "hoop, holler, and hurrah" to find your way back out. And I was warned to beware of the wild boars, known locally as "piney-woods rooters." I was somewhat skeptical of many of his tales, but I did wonder about my abilities to navigate through, or even around, the infamous and nearly impenetrable titi swamps. Two species of small deciduous trees (black and white titi) and other shrubs and vines make this thickly vegetated swamp challenging to traverse.

Dr. Dan missed no chance to get out into the field, and his true colors, adventurous nature, and woodsmanship showed whenever he could escape the office. But leave him in the dreaded office for too long (mounds of paperwork come with any academic or agency position), and he became like a cornered panther, ready to lash out. The grad students often gathered at the coffee pot, and I was once expounding on some subject related to our ethical responsibilities as biologists. Under other circumstances, Dr. Dan might have agreed with me. But a prolonged office confinement had caused him to morph, and the students scattered like spooked deer when he suddenly appeared, glared at me, and growled, "You're not a biologist;

you're just a peon." My compadres never did let me live that one down.

Before I was actually turned loose to begin my indigo snake distribution study, I traveled to southern Georgia with Dr. Dan and his primary field technician, Joe, to "learn the ropes" (how to hunt for indigo snakes in their natural habitat). At the time, a companion study to mine involved the ecology of the indigo snake, and the research was being conducted on a seven thousand–acre site near Tifton, Georgia. The land was owned by a timber company but was under a cooperative agreement with Auburn University. Much to the dismay of the local populace, the vast acreage was posted as a "snake sanctuary."

It was in this sanctuary that I initially gained firsthand knowledge about sandhill flora and fauna. I learned to recognize a key indicator of sandhills, turkey oak, so named because the leaf is shaped like the foot of a turkey. Planted slash pine had replaced the natural longleaf pine in most of the sanctuary, but the habitat was so well managed via thinning and prescribed fire (making it open and grassy) that the area supported robust populations of both gopher tortoises and indigo snakes. Early on, I seemed to have an innate propensity for walking up to, but thankfully not treading on, cryptic eastern diamondback rattlesnakes. Dr. Dan was delighted that I could find his much-admired diamondbacks and promptly deemed me a "white witch woman," which I took as a compliment.

Dr. Dan and Joe delighted in playing good-natured practical jokes on me during our field forays and even when we returned to civilization. We would generally split up when snake hunting so that we could cover more area. On several occasions, I would hear snickers and stifled laughter before I heard the truck doors slamming. I guess they wanted to see if I would freak out when I heard the truck drive away. I surely wasn't about to give them the satisfaction of knowing that being alone bothered me (which it didn't unless I had to hitchhike my way back to Tifton), so I just moseyed along until I eventually heard the truck return. Of course, they professed ignorance and said that they didn't know where I was and had therefore gone looking for me.

Their premier practical joke took place in a little café in

Ashburn. Even though it was technically still spring, the Georgia heat and humidity were making their presence known, and after a morning of stomping around the sandy lands, we were grungy. After we placed our lunch orders with the gray-haired, grandmotherly waitress, I retired to the restroom to try to wash off some of the sweat and field dirt. When I returned to the table, I was mystified about why the waitress, who had been South-Georgia-friendly earlier, was now staring and giving me the evil eye. My hands and sunburned face were certainly cleaner than when I had placed my order; so what had changed? After a while, she came over with a quizzical look on her face and addressed me and Joe: "You two *really* are married, aren't you?"

That comment surely perplexed me, and after some awkward back-and-forth exchange, I learned that while I was cleaning up, she had asked if I was Joe's wife. With their best feigned sympathetic (and lying) faces, Dr. Dan and Joe had told her that they didn't really know me: I was just some unfortunate, poor white trash that they had picked up out on the interstate. As soon as she finally realized that this was a joke, we all had a good laugh. But I still think she thought we were weird, especially when it was revealed that our sweaty appearances were because we had been snake hunting.

One of the premier gatherings for wildlife biologists in our area was the annual Southeastern Wildlife Conference. Apparently, in its early days, the evenings met a different definition of *wild* life. Thankfully, by the time I attended my first conference in 1978, things had settled down quite a bit. Nevertheless, as I sat with the other Auburn wildlife grad students at the evening's banquet in the grand Homestead Hotel in Hot Springs, Virginia, a somewhat inebriated Dr. Dan kept bringing male colleagues over to meet me. At first, I thought it was good of him to introduce me to other wildlife professionals. But as the night wore on and the beer flowed freely, his words began to slur and he almost dragged one fellow over to me and said, "This is my grad student: dance with her!" At that point, I was mortified. I reminded Dr. Dan that I was married and that this was not the way I wanted to meet professional colleagues. He glared at me for a moment and retorted, "Well, you don't have to be insultin' about it." As he stomped away, my more-seasoned cronies shook their heads and told me I had better hope he was drunk enough to forget what I said; otherwise, my research assistantship was history. Fortunately for me, he had completely forgotten about the entire incident by the next day. Although I was embarrassed to be singled out at the time, I now think that this was Dr. Dan's albeit misguided, booze-infused, way of initiating me into his world of wildlife conferences.

One of my favorite Dr. Dan stories originated from the man himself and is another example of his ability to laugh at himself and at life's quirkiness. In addition to the indigo snake research, Dr. Dan was engaged in a study of wild turkeys at a remote hunting camp in Choctaw Bluff, Alabama. In fact, he was so well known as a turkey biologist that I later discovered his name in the old notes from my undergrad wildlife management class. At the time of this particular study, southern Alabama north of Mobile harkened back to an earlier era. Generations of Alabamians had hunted in these backwoods, and the hunting camp reflected that history. Dr. Dan and his students at the time would conduct research and stay at the camp. One of the older, wizened, black hunting camp employees repeatedly heard the students address their professor as "Doctor." Finally, he inquired of Dr. Dan, "Just what kind of doctor *is* you?" When Dr. Dan explained that he was a PhD, not an MD, the old man shook his head knowingly and replied, "Um-hum, you one of them doctors that can't do nobody no good."

Dr. Dan and Ruth always had a delightful menagerie of domestic and wild critters at their home. The latter were often refugees who were being rehabilitated or needed to mature before being released. I always enjoyed going over to their house to interact with the animals. Although I never met him (he was before my time), one of their most noteworthy "family additions" was Bobo, the black buzzard. A wildlife student found Bobo in an abandoned house when the bird was only three weeks old. Bobo quickly assimilated into Dr. Dan's household, learning to shake "hands" with his beak; more or less fetch balls and sticks (he would go after them but didn't bring them back); and enjoyed playing a game of chase with the family bulldog.

My favorite Bobo story involved a day when Ruth was going to run a few errands, and Dr. Dan insisted that she take Bobo with her: "Bobo really wants to go with you." Against her better judgment,

Ruth put Bobo in his position of choice, the top of the front seat near Ruth's shoulder. Bobo had apparently been recently fed his liver or kidney that morning, and sometimes his food was spiced with his favored garlic and onions. As Ruth drove down Auburn's tree-lined streets, Bobo seemed to be enjoying the ride . . . until he wasn't having so much fun and regurgitated his morning's meal all over Ruth. Needless to say, vulture vomit is decidedly unappealing. She had no choice but to promptly turn around and go back to the house. As Dr. Dan came out to see why Bobo was being short-changed on his joy ride, Ruth glared at him and said, "You go straight to hell!"

As a pillar of the society in Auburn, Ruth had ties to many organizations. At the time, she was involved with the Lee County Historical Society museum in the tiny town of Loachapoka, just west of Auburn. Loachapoka was best known for its annual syrup-sopping day. The town's museum edifice was decidedly from a bygone era, and one day, Ruth called us at the Wildlife Unit to ask if someone could please come remove a large gray rat snake that had taken up residence in the ladies' room toilet and was apparently terrifying the older, genteel ladies who had need of this bathroom.

Before someone had a heart attack, I (as the unit's snake lady) was dispatched to perform this exorcism of sorts. I must admit to being a bit taken aback when I saw the snake happily curled within the confines of the toilet bowl. As I rolled up my sleeve and slowly reached down to grasp the snake, it decided that perhaps a safer alternative existed farther down in the bowels (if you'll pardon the pun) of the ancient toilet system. And so began a prolonged tug of war, with me sprawled on the bathroom floor and my arm way down in the toilet. I sure didn't want to injure the snake so that part of it remained in the plumbing; therefore, infinite patience and a persistent, but gentle, give-and-take motion were required. Amazingly, and thankfully, I was eventually able to extract the recalcitrant reptile from its rather bizarre refugium.

The Wild Bunch

Although I was technically part of the Wildlife Department at Auburn, the Co-op Unit was its own separate lair down the hill from the main academic buildings. And not unexpectedly, my fellow grad students and technicians at the unit were colorful in their own right and became my "peeps" (to use a slang term that didn't exist back then). More than three decades later, some are still close friends, and we've shared many an adventure in the distant and recent past. When I first arrived in Auburn, I met Joe and Doug. Joe was the primary technician on the indigo snake ecology project. He was an excellent field biologist who heralded from northern Alabama. His Southern accent is still strong to this day, and we once had an ongoing argument over a word that he swore existed: *heldt*, as in "I heldt it in my hand." At the time, I was posting a word of the week to expand our vocabulary horizons, and Joe added *heldt* to the list. He almost had me convinced that it was an English word—and I reckon it is in Southernese.

If there was a leader and mentor grad student, it was Doug, who was working on his PhD researching bobcats. He was razor sharp in intellect and had a finely honed and understated sense of humor. After we had watched a film about a woman biologist working on giant otters who coincidently changed spark plugs in her boat (a big deal, apparently), Doug gave me a plaque: The Joan "Husky Woman" Diemer Spark Plug Award, with three spark plugs wired to the wood. The "Husky Woman" nickname came from Dr. Dan calling me "husky" because I could dig out gopher tortoise burrows in pursuit of indigo snakes.

Another very Southern and frequently hilarious grad student, Danny, was working on his PhD on wild turkeys. He was big in size and heart—and I once had to help extract him from quicksand as we meandered along an Alabama tannic-stained creek in pursuit of who knows what. I always got a kick out of his telling his younger wife (who was both feisty and prone to tears), "I didn't take you to raise!" I lost track of Danny a few years after I graduated, and, heartbreakingly, Doug's career and life were cut short by brain cancer. But I will always remember these amazing men who enhanced both my educational experience and my life at Auburn.

Two other members of my new family arrived shortly after I did: Donna and Mark. They were hired as YACC's: Young Adult Conservation Corps workers. Donna became my right-hand woman and trusted field tech, who accompanied me on field forays into South

Georgia. We both had grown up in Maryland and bonded immediately. She was, and still is, a petite beauty; she and I could always share stories about the quirkiness of life. You know you are good friends when you have to pluck a tick from someone's posterior, and I did that for Donna (though I'm sure Dr. Dan would have been glad to do so). And how special is it that she fell in love with, and married, Joe?

The other YACC, Mark, also remains a friend after all these years. He and I have had many misadventures camping and backpacking in the West. Back during the indigo snake study, Mark, Donna, and I spent many days traveling through South Georgia in pursuit of information about indigo snakes. We sweated profusely in the oppressive Georgia heat and humidity as we hunted for indigos. At times, the moisture-laden air seemed to wrap around us and slow our steps. I reckon "only mad dogs and Englishmen," and field biologists, venture out in the noonday sun.

After one especially long day in the field, we washed off the Georgia dirt and retreated to the cool, dark comfort of an antique-filled eatery in Tifton. The owner, an ancient, genteel man (who must have had poor eyesight, right?), somehow thought that I was the mother and that Mark and Donna were my offspring. I was under thirty, and they were about four or five years younger, but this elderly Southern gentleman was convinced that we were a family. To this day, I still call Mark my "son."

After another typically sweat-inducing day in the field, Donna and I got cleaned up and were anxious to go eat but couldn't get Mark to answer the phone in his motel room. So we strolled over to his room. Donna knocked. Still no answer, so she peered through a crack in the curtains into the dark recesses of the room and jumped back when her eyes spied Mark's bare behind as he was getting dressed after emerging from a long shower. Well, I guess in a way, we were family after all.

My Inaugural Wildlife Study

Now firmly immersed in the colorful culture of the South and armed with just enough initial knowledge of local wildlife and their habitats, I was ready to embark on my master's research project. The objectives of the study were basically twofold: to survey the distribution and current status of the eastern indigo snake in Georgia and to characterize and delineate its habitat in the state.

My first task was to create and send out questionnaires to potentially knowledgeable persons to solicit information regarding indigo snake distribution and status: 373 questionnaires went to herpetologists (scientists who study reptiles and amphibians), wildlife biologists, museum curators, state conservation officers (game wardens), local landowners, and amateur naturalists. For nonscientists, the questionnaire included a description of the indigo snake and an inquiry about the person's ability to identify this species. I checked the mail each day, awaiting the return of the questionnaires. Eventually, I received 159 completed questionnaires, with 62 containing reports of indigo snake sightings. This, I was to learn, was only the beginning, my entry point into a previously untapped resource.

Why? you might ask. Because as I began to follow up on the questionnaires by conducting interviews and field surveys with those individuals who indicated that they had indeed seen an indigo snake, the names of many additional contacts were provided to me. Sometimes, I swear, the list of names seemed endless, but I did my best to track down each and every potential source of information. And it was this tenacious pursuit of people and the subsequent interviews to glean details of the sightings that not only provided my desired indigo snake distribution data but also yielded fodder for some intriguing stories of my South Georgia adventures.

In many cases, I visited the actual locality of the sighting with the person so that I could complete a vegetation analysis (habitat type, major plant species within fifty feet, distance to water, and whether gopher tortoise burrows were present). If the site seemed lucrative and the interviewee agreeable, I would schedule snake-hunting forays. In so doing, I was temporarily transported into the daily lives of the interviewees and was frequently offered a seeming cornucopia of Southern cuisine. Note: Refusing food in South Georgia is akin to insulting both the family and their culture; I *did not* refuse food. And as crazy as it might sound, I firmly believe that my partaking of these tasty morsels was often a key to my success in establishing working relationships and procuring information. Especially in the

South, the preparation of food and the subsequent consumption of it is a time for storytelling and relating events old and new.

In one case, Dr. Dan's wife, Ruth, and I shelled beans as we heard about indigos and rattlesnakes. Because folks knew I was interested in wildlife, dinner-table conversation might revolve around "black panthers" or other questionable beasts roaming the Georgia piney woods. I was particularly fascinated by the cultural and historical tales. For example, one interviewee was apparently bitten by a venomous cottonmouth (water moccasin) as a child; his family gave him booze and watermelon. Surprisingly, he survived. Not so surprisingly, he had permanent tissue damage.

My research involved a multipronged approach to gathering data. In addition to the questionnaires, follow-up interviews, site visits, habitat surveys, and directed snake hunting, I attended six rattlesnake roundups over two years to interview (and in a few cases, befriend) the snake hunters who might encounter indigos when targeting venomous snakes. Both the eastern diamondback rattlesnake and the indigo snake are found in sandhill habitat and use gopher tortoise burrows, so these local snake hunters were prime potential sources of information. The Georgia roundups at that time took place in Claxton (famous for fruitcakes), Fitzgerald, and Whigham. These events were a boon to the towns' local economies and often had evolved after someone had been bitten by a rattlesnake. The mentality of ridding the countryside of the dread serpent was pervasive, but over time some of the roundups allowed Dr. Dan to add information to the roundup programs about the ecology and conservation needs of indigo snakes and gopher tortoises.

The gathered snakes were sold to venom facilities during the late 1970s. At other times in the history of these roundups, snakes might be killed and sold for meat and hide. The pharmaceutical uses of venom drove the need to keep the snakes alive to be milked. However, unbeknown to the local populace, some snake buyers turned the serpents loose again after an initial milking. Most roundups have been either been terminated or modified to not have such a deleterious effect on the environment and the local fauna.

To drive rattlesnakes from their subterranean retreats in winter and early spring, snake hunters poured gasoline through a hose into gopher tortoise burrows, killing indigos and certainly not helping the resident tortoises. Although I was hardly thrilled over the mass gathering of rattlesnakes for roundups, I had to be cautious not to transmit an eco-freak mentality when entering the inner sanctum of the snake ring. At first, the snake hunters were convinced I was from the Sierra Club and had come to stop the roundups. My gender may have initially fostered that belief, but I also think that it allowed me liberties that a male biologist would not have had. It took a lot of talking and persistence to convince them that I was just a graduate student gathering information for my thesis.

The exchange of trust went both ways: Most of the snake hunters came to realize that I wasn't there to judge them, and I learned that these weren't depraved, slavering hicks. You see, Dr. Dan had required all of us on the indigo snake project to read Harry Crews's *A Feast of Snakes*, about a fictitious town's rattlesnake rodeo and the morally depraved characters who inhabited Mystic, Georgia. In contrast to Harry Crews's dark, grotesque, and violent depiction of South Georgia roundups and the local populace, my experience convinced me of the innate kindness and helpfulness of these rural folk. They may not have always completely understood why this tall blonde with the somewhat Yankee-fied, somewhat western accent was chasing snakes (I was often initially asked, "Where you from, gal?"), but they accepted my seeming "strangeness" and chose to assist me.

Eye in the Sky

There was still another essential element to my indigo snake study: delineating the sandhill habitat in ninety-four Georgia Coastal Plain counties. Fortunately, I have always absolutely loved maps (a genetic or bestowed gift from my father), because I certainly spent countless hours examining soils and vegetation maps for each county. But Dr. Dan and I wanted to go one step further; we employed an early version of Landsat satellite imagery. This was relatively new and exciting technology in the 1970s, and our methods were loosely based on that used to evaluate grizzly bear habitat over vast areas of northwestern Montana. In fact, although I initially questioned why

Dr. Dan pulled me from the field to go to a Remote Sensing Symposium in South Dakota, I was extremely grateful when I was seated at the lunch table next to one of my wildlife biologist heroes: grizzly bear researcher extraordinaire, John Craighead. I was honored to be able to discuss and compare his use of Landsat imagery with my own, to better understand and benefit two obviously very different species.

Basically, the imagery that we used in our studies was derived from the earth-orbiting satellite Landsat 2. As the satellite passed over a 115-mile swath of land, known as a "scene," the reflectance values from one-acre units, called "pixels," were detected. For the indigo snake study, I used winter data because the absence of leaves on the scrub oaks during that season enhanced the reflectance value (known as a "signature") of the sandhills. I viewed the data via a computer in Atlanta. PCs didn't exist yet, and, thankfully, I had computer geeks and remote-sensing specialists to assist me. Because the data were portrayed in an infrared mode, growing vegetation appeared red, not green; river swamps were blue-purple; cleared farm fields were a pale greenish-white; and the targeted sandhills were light tan-brown.

After months of orienting the data on a large state highway map and then photographing each frame (ninety-eight frames/scene) on a video screen, the final product was 212 slides (each covering nine by twelve miles) that could be projected onto a large screen to look for sandhill habitat. It was tedious but fascinating work, and employing new, space-age technology proved helpful in eventually getting my indigo snake distribution study results published in a scientific journal: a raison d'être and major goal for research scientists. Of course, the Geographic Information System (GIS), Google Earth, and other modern tech-mapping programs would have made this undertaking much easier, but one uses what one has at the time.

Snake Hunting 101

When I wasn't in Hot-lanta (apt name for Atlanta) working on habitat delineation or immersed in classes at Auburn, I was on the road in gorgeous Georgia. I calculated that I drove around twenty-three thousand miles over the two-year study. One humorous entry from my field journal read "evening pleasant, uneventful except for naked guy on I-75!" Although I occasionally did use a federal vehicle, most of my travel was via my little red Chevy Chevette. Using my own vehicle elicited less concern from the locals, while driving up in an official vehicle could be off-putting. And the diminutive Chevette had an impressive turning radius, essential for making quick U-turns when a DOR (dead-on-road) snake was sighted. I don't know how many times poor Donna hollered "Geez!" as I spun the car around on some backcountry road.

The reason we had to check dead snakes was that indigo records were still valid even if the snake had unfortunately become vulture fare. Simply driving by a squashed snake wasn't enough; we had to get out and pick through the remains like white witch women indeed. Even *half-mortified* (my snake hunter best bud Amos's term for not so recently deceased critters) snakes had to be examined, especially if they were black. Sometimes, it boiled down to checking the anal plate, the scale directly above the waste-elimination orifice, to see if it was divided (meaning it was likely a black racer) or undivided (which could mean it was a small indigo). Really large black snakes were typically indigos—but I encountered many more small or medium-sized DOR black racers (not necessarily surprising since indigos are imperiled).

Especially if I was alone, local folks thought I was either crazy or needed auto assistance. In one case, I went back twice to verify that a DOR black racer was not an indigo. I was on a back road that wound up along a bluff overlooking the relatively wide Ocmulgee River when a man stopped to ask if I needed help. I'm sure he wondered why I was down in the road looking at a half-mortified snake if I was merely "just checking my car." I don't recall if I told him what I was actually doing, but I doubt he would have understood anyway. In another similar incident, my field notes reflected how strange my behavior must have looked to black field hands, who ceased their activity and just couldn't seem to stop staring at me. To those onlookers, I must have appeared as some bizarre version of *La Dame Blanche*, better known in the Deep South as a "haint," a ghostly apparition or supernatural spirit. Although my field notes generally contained biological or sociocultural information related to my study, I found another entertaining entry written after I must have stopped to examine a dead snake. "Yuk: ate melted breakfast bar and licked fingers. Had rat snake blood on it! I'll probably get parasites: Yuk!"

Believe it or not, there is a method to the seeming madness of searching for elusive indigo snakes. They are decidedly uncommon, so timing, habitat (which I was researching), and just pure luck are elements of the search. Late-winter and early-spring months (especially February to April) were optimal to hunt for indigos basking outside gopher tortoise burrows on warmer days. Remember that snakes are cold-blooded, meaning that they don't thermoregulate like we do; instead, their temperature is determined by the ambient temperature around them. That's one reason indigos like the deep recesses of tortoise burrows. I have always maintained that I can empathize with reptiles because I too have (as one of my buddies in Florida used to say) "a narrow zone of thermo-neutrality."

Even if I didn't find an indigo snake, its shed skin was distinctive and could serve as positive evidence of occurrence. Although snake tracks in loose sand are not species-specific, they could indicate relative body size and serve as indicators of possible specimens. Large, heavy-bodied rattlesnakes may have straighter tracks, while indigo tracks might be relatively more sinuous. But a shed skin, a confirmed *sure enough* (a Southern term) sighting, and a snake in hand were the only definitive data. On rare occasions, after my colleagues and I used a hose to listen down in tortoise burrows that had large snake tracks in the vicinity, we might decide to undertake the arduous job of excavating a burrow if we really believed that an indigo was present. As spring transitioned into the long, hot summer, the searches around tortoise burrows tended to be less profitable because the indigos moved into the sandhill/river swamp edges (called *ecotones*).

Searching along river swamps was truly akin to looking for a needle in a (wet) haystack. I was more likely to encounter cottonmouths (a venomous species) or other water snakes rather than indigos. On one occasion in a coastal area, I climbed a huge mound at dusk and then spied a snake on the road. It was likely either a nonvenomous water snake or a small cottonmouth. I carefully approached the snake and crouched down to try to get a better look, but it was too dark and I sure wasn't about to grab it without a positive ID.

Snake Hunters Extraordinaire

Although I interviewed numerous individuals during my study, dead-ends, in terms of useful information, were not uncommon. However, 111 additional persons, besides the original 62 who responded to the questionnaire, reported indigo snake sightings. Because some reports were more solid than others, I devised a validity classification that considered both professional and individual abilities. Biological aspects notwithstanding, many of those interviewees were intriguing in their own right. However, there were five whom I got to know better and who stood out from the rest. I was, and still am, extremely appreciative for the information and field hours that they so generously donated.

All of these men were snake hunters who enjoyed getting out in the woods. Most pursued primarily eastern diamondback rattlesnakes, but one was a devoted naturalist who loved all snakes. This colorful character would have considered himself a "swamp man," and he was, and still is, known as "Okefenokee Joe." The other four might (or might not) have been proud to be *Georgia crackers*. This was not a snack food. Instead, these rural folks were allegedly so named because during the pioneering days, they would drive their livestock through the sandy lands and along trails that wound beneath bowers of gigantic live oaks, all the while cracking their whips. An older and alternative etymology heralds from the Elizabethan era when the term was used to describe a braggart, a wisecracker. These Southern backwoods were historically settled by tough and resourceful folks who came from the Celtic-influenced regions of Britain. Later, in Florida, I would encounter many crackers who had deep roots in the Sunshine State's sandy soil. I've always considered being a cracker as a positive attribute, in that many descendants of the original Georgia and Florida crackers still have deep ties to the land. Comedian Jeff Foxworthy later made being a redneck a source of pride and good-natured humor, and I once noted that an interviewee reminded me of a "redneck Humphrey Bogart." That said, I still feel that *cracker* is a more historic and less pejorative term.

In doing research for this book via my field journals and published scientific literature, the internet, and of course my internal memory bank, I found that my old amigo Okefenokee Joe had become somewhat of a phenomenon. I knew that he had roots in country music, but he's re-created himself through his evocative songs about nature, especially the Okefenokee Swamp. And I would highly recommend his 1990 video *Swampwise*, available online through Georgia Public Broadcasting media. Okefenokee Joe, a.k.a. Dick, is a poet, balladeer, troubadour, educator, naturalist extraordinaire, and ardent environmentalist. He provided me with a wealth of knowledge about the history of indigo snakes in Georgia and the many threats they faced.

At the time of my study, he was living in the Okefenokee Swamp with his younger girlfriend Cindy (who later became his wife). He was working as a curator and educator for the swamp park in Waycross and had apparently caused a bit of an uproar when Cindy, a local prominent doctor's daughter, fell for his considerable charms and left home to "shack up" with him in his backwoods abode. They seemed quite happy when I stayed with them one memorable night.

Presumably because Okefenokee Joe was a well-known celebrity and did outstanding snake programs (at the time, he had a cottonmouth snake named "Pussycat"), a male reporter/writer was also visiting that night. The swamp house was rustic and tiny, with lots of wild and domestic critters, some inside and most roaming about outside. Snake cages lined parts of the house. Okefenokee Joe set up a cot for me in the living room, with another cot for the writer nearby. Sleeping that close to someone you've just met doesn't necessarily make for a great night's sleep anyway, but the awkwardness was increased tenfold when we heard cooing (not from doves) and bed squeaking and giggling emanating from the main bedroom. Obviously, there was some romancing going on in that otherwise very quiet swamp cabin.

One of the first snake hunters that I befriended was J. C., a pillar of his community, who resided in what would turn out to be the heart of Georgia's indigo snake territory. J. C. was a real find indeed, in terms of how much he contributed to my study. I don't recall the names of his family members, but for some reason, I remember that his dog was Flossie (probably because you had to keep an eye on Flossie, especially when you first arrived and she was in farm-guard mode). J. C. had many indigo snake sightings of his own, and I also hunted with him to add to my database.

One early spring day just prior to a rattlesnake roundup, J. C. and I captured both a huge eastern diamondback rattlesnake and a large male indigo snake. The diamondback was crawling just outside a gopher tortoise burrow; and about fifty feet away, the indigo was at a collapsed burrow with a tiny opening. My field notes reflected this rare and fortuitous find: "What a thrill!" J. C. took the rattlesnake to the roundup, and it placed third in terms of its size and weight, five and a half feet and nine and a half pounds. Because we hadn't expected to necessarily find a rattlesnake, we didn't have a snake box (safer when catching venomous snakes). Instead, we were using the typical pillowcase snake sacks, and I distinctly recall some understandable angst related to the massive rattlesnake head going past my hands as I held open the sack and J. C. eased the viper down into the bottom.

On another memorable nonsnaky occasion, J .C. called me and said that he was accompanying a livestock provider to Auburn and would be assisting with a collegiate rodeo there. Would I like to come see the Brahma bulls? Yes, I would. I was able to get up close and personal as the young college cowboys secured their hats and eased onto their assigned bulls in the chute. Although the bulls were relatively young too, they were still gigantic and also inexperienced as rodeo stock. One bull seemed especially agitated, and I literally had to leap off the fence when he climbed (who knew they could do this?) up the fence, momentarily loomed above me, and jumped too close for comfort as he made his escape. It took the cowboys quite a while to contain and herd that bull back to a holding pen. Their horses had little experience with refrigerator-sized bovines and therefore wanted nothing to do with the escapee.

Two other snake hunters who went well beyond the call of duty were Mel and Rodney. They often hunted together, and I got to know Rodney via Mel. Mel was a jovial fellow who worked as a guard at the state prison in Reidsville. On several occasions, I would meet him as he was getting off work. It was always eerie to hear the sound of the gates clank shut as I entered the visitor area of the prison. Mel and his wife, Cathy, were fine folks indeed and extended many kindnesses to Donna and me. Of course, in typical South Georgia style, they didn't let us go hungry.

I never really got to know Rodney as well, but Donna and I had a standing joke that you didn't want to turn your back for too long when snake hunting with Rodney. I feel sure he was harmless, but he had a habit of turning up a bit too close for comfort at unexpected times. Despite being a close-stander, he was good at helping us find indigos, so much so that one male snake that we took back to the Auburn Lab as a breeder was named Rodney in his honor. On one bonanza day, Rodney (the person) and Dr. Dan flushed a relatively large female indigo snake (males are typically longer than females) from where she was hiding in a brush pile, and I grabbed her. About one hundred feet away, they removed leaves from a hole to find a robust male indigo trying to take refuge, and we did some quick digging to capture the big guy. (I think love was in the air.) Then, we put a listening hose down a tortoise burrow located between the two capture locations and heard the hissing of a snake. We dug halfway to China, seemingly, and found another female indigo. Once again, my field journal succinctly reflected this providential and pretty amazing capture event: "What a day!"

If I had a favorite snake hunter, it was Amos. I never really knew my grandparents, and Amos and his wonderful wife, Jean, became surrogate G-parents. They basically adopted me and Donna, and when they met Joe, Donna's eventual husband, Amos deemed him "a prince of a young man." I stayed with them on many occasions and felt like I was coming home when I saw their pale green Claxton abode. I loved them dearly, and I know they loved me.

I first met Amos in the Claxton rattlesnake roundup's inner ring, where typically only snake hunters were allowed. As I've said, I had gained some semblance of trust and enough "field cred" (a backwoods version of street credibility) to be allowed into the ring to interview the hunters. Like the other top rattlesnake hunters, Amos had seen a number of indigos over the years. I later spent many enjoyable hours snake hunting with Amos. We found shed skins indicating indigo presence, and on one occasion, Amos caught an indigo for me and held it for a day or two (a bit risky considering he didn't have a permit) until I could get over to Claxton.

Amos and Jean's kindness truly knew no bounds: They fed me huge quantities of often Southern-fried food until I nearly foundered in the Georgia heat. I even made mention of a particularly tasty gumbo in my field journal. And because he and Jean worried about my being thin, they couldn't resist giving me edible goodies to take on my travels.

Amos also made me a hard-sided, wooden box so that I could safely transport snakes, especially venomous ones. Being part of the extended family, I came to know their daughter and her family, as well as Jean's sister, Nannie-Byrd, who was the mother of a well-known state politician. One time when Donna and I were supposed to stay with Amos and Jean the night before the annual Claxton rattlesnake roundup, we saw a note on their front door indicating that he and Jean had gone to the ER after a refrigerator fell on Amos and injured his arms. As instructed in the note, we found Jean's sister's house after stumbling around in the dark and almost getting eaten by a Cujo-esque, scarily big dog. Thankfully, Nannie-Byrd took the wayfaring snake gals under her wing, provided lodging, and treated us to a much-appreciated and delicious dinner. At some point during my indigo snake years, I even got a VIP tour of the Claxton Fruitcake Company. Needless to say, I had plenty of Claxton fruitcakes to give as gifts. I kept up with Amos and Jean for quite a while after my indigo snake study ended, and they will always hold a very special place in my heart.

In addition to the five snake hunters who contributed so much, others were noteworthy for various reasons. In one case, Donna and I tracked down a reference named Wendell, and he agreed to show us where he had seen indigo snakes in the recent past. As we drove up to the rendezvous point, Donna expressed concern about the shockingly rough-looking dude who awaited our arrival. "Oh, I'm sure he's fine; quit worrying," I replied. After all, he had been referred to us by someone who seemed reputable, so he must be okay, right? Donna's fears subsided when Wendell acted like a complete gentleman and, because she had a cold, he even piggybacked her across a creek so she wouldn't get wet. The next day, we just happened to meet with a local Georgia State Patrol supervisor, who coincidentally was also a snake hunter. When he heard we had spent the entire previous day with Wendell, his eyebrows raised a bit, and he told us that Wendell had a criminal record for burglary and car theft. Needless to say, I had to eat some serious crow for dismissing Donna's initial concerns.

Snake Ladies

Tracking down persons who might have seen indigo snakes was often quite challenging. Remember that answering machines were rare and cell phones nonexistent in the late 1970s. Consequently, I either had to keep calling folks from my motels in the morning and evening, or I had to find and use pay phones. In one case, I went to five pay phone booths: Four didn't work and one was occupied. I finally had to go to a gas station and ask to use the phone. And I frequently had to leave messages with other family members, especially wives, who probably wondered why in the world this woman was calling their loved one to inquire about snakes. My field journal summed this up: "Where does persistence grade into *nuisance*?"

I was both an anomaly and enigma to many of the South Georgia folks. Women field biologists were rare enough, but one who searched for snakes fell into the bizarre range. Once, as I drove away after asking directions at a gas station, I heard a local bystander inquire, "*Who* works with snakes?" On another occasion, a bona fide redneck looked me over and quipped that there were two things that he had a horror of: snakes and wild women. And if one woman chasing snakes seemed strange, two women traveling around in pursuit of snakes was downright mind-boggling.

Nevertheless, our fame sometimes preceded us. One rainy day when we couldn't drum up an interview, Donna and I decided to go the Agrirama (now known as the museum and historic village) at an agricultural college in Tifton. As we walked in, the woman who was giving a tour stopped in midsentence and exclaimed, "My, my;

it's the snake ladies!" We couldn't help wondering if we looked like Medusa, with snakes emanating from our hair.

Strange Men

Not surprisingly, some interviewees just had to show off to one degree or another. These were not necessarily the tried-and-true snake hunters. In one case, a landowner was determined that we should look at potential indigo snake habitat via his dune buggy. And of course, we had to go sliding down steep sandhills and even did a quick jaunt on a railroad track in a "Look ma, no hands" mode: He didn't steer! In another case, I was interviewing a local wildlife officer. As we rode along the back roads in his state vehicle, he got a call regarding a dead-on-road deer. Suddenly, he accelerated to seventy miles per hour on these dirt roads, to get to a dead deer? I tightened my seat belt and said a prayer that we wouldn't careen into the trees.

Fortunately, nearly all the men I interviewed behaved like gentlemen even if they thought I was unusual. That said, there were exceptions. Two of these exceptions involved men in law enforcement. I guess I shouldn't have been too surprised. The first was a wise-guy wildlife officer, who looked at me as we finished surveying potential habitat and asked what I would do if he left me locked inside this fenced property. I wasn't about to let him scare me; I looked him directly in the eyes and firmly declared that I would hike out and go straight to his supervisor. That was the end of that, but it did make the drive back a bit awkward.

In an even more atypical instance, a black federal wildlife officer (rare then, especially in the South) asked me if I would go out with him when he came back to the area where I was working. It took me a minute to catch his drift (we were both married, after all), but, again, I wasn't going to let on that he rattled me, so I replied, "Of course, we can go out in the field and look for snakes." He was neither a woodsy nor a snake person, so that shut him up.

He also caused some problems for me at the first Whigham rattlesnake roundup I attended, when he went undercover and tried to dress like a local by wearing overalls. He was looking to see if anyone was bringing in indigo snakes illegally, but by walking up and chatting conspiratorially with me on several occasions, he undoubtedly compromised my credibility. His ruse definitely didn't work. As an urban black man with nary a trace of a Southern accent, he wasn't about to be accepted as one of the locals.

The most bizarre and unsettling incident involved neither a stupid comment nor a direct come-on. Even today, all these years later, I cannot figure out whether this was an extremely unusual coincidence or something much more potentially insidious. I do feel that I must have had guardian angels looking over me throughout my career, for I certainly encountered a number of dicey scenarios and survived many misadventures. This strange incident started out harmless enough. In the early afternoon, I was interviewing the owner of a fertilizer plant about possible indigo snake sightings and potential habitat. A number of other men (whom I assumed were locals) were standing around. It's important to note that this fertilizer plant was in extreme southwestern Georgia, more than eighty miles as the crow flies from my evening destination in my frequent home base of Tifton. And another salient fact that plays into the strangeness factor was that I hardly took the crow's route to Tifton.

After leaving the fertilizer plant, I went to scope out some of the suggested potential habitat, only to find that high water prevented my entrance into the desired area. I had to circumnavigate the site only to be stopped again by a locked gate. After more meandering on back roads that almost took me in a big circle, I headed farther east, stopped to make a call, and then interviewed a crop duster at his hangar. Finally, I picked up a more direct road and drove a final hour to Tifton, arriving after 6:00 p.m. I grabbed a bite to eat at Pizza Hut and proceeded to my motel.

My field journal, which generally covered primarily biological or cultural insights directly related to the snake study, deviated to note: "Weird: the guy I met at Mr. H's (fertilizer)—thought he was local—ended up behind me and pulled into motel behind me. Seemed nice, distinguished looking. Asked if I wanted a drink before studying." I had apparently mentioned something about needing to study for a big test back at Auburn.

My journal went on, "Then he ended up in room next to me:

strange coincidence!" Truly, I can't imagine that this "distinguished-looking" man followed me all over southwestern Georgia as I perambulated on back roads, stopped to do an interview at a professional business venue, and then made my way to Tifton many hours later. I suppose it's possible that he saw my car on the more mainstream highway and followed me from that point, or perhaps he followed me from Pizza Hut. But how and why did he end up in the room next to mine? Needless to say, I did not venture back out that night and engaged every lock on that motel door. I stayed up until 1:00 a.m. and noted that it was "depressing to study in a motel room." But I suspect that it wasn't just that I had to study that night. I'm sure that I remained on edge after the evening's earlier perplexing and unnerving incident.

A Night in Jail

Of all my many misadventures in South Georgia, the night after my first Fitzgerald rattlesnake roundup ranks right up there as one of the most head-shakingly bizarre. I had spent St. Paddy's Day (March 17, 1979) at the roundup, interviewing local snake hunters and the colorful guy who was actually purchasing the snakes for venom extraction. My field journal noted the following about Jack, the toxin researcher: "Is quite a character; very knowledgeable biochemically; B.S.'s a bit on some things but knows his snakes."

Jack and his colleagues were giving snake talks and had brought what some folks might call a "warlock's brew" of live venomous snakes: both eastern and western diamondback rattlers, prairie and rock rattlers, a canebrake rattler, a banded krait, monocled and spectacled cobras, and a young (eight- to ten-foot-long) king cobra. At the end of a long (but certainly interesting) day of gathering information about indigo snake sightings and potential habitat, one of the roundup organizers somewhat sheepishly approached me and asked if I would like to have an indigo snake that had been caught earlier that day.

Now these local snake hunters knew it was illegal to have an indigo, but word may have gotten out that the Auburn snake lady was looking for these blue-black beauties. Or perhaps someone thought it could be sold at the roundup; who knows? As the snake hunters milled around in the inner ring, seemingly a bit uncertain how to proceed, I marched in and said I would take the snake, and before anyone else could do anything, I took it. Fortunately, I did have some backup in the form of two Georgia Department of Natural Resources biologists. (You may recall that Georgia DNR was funding the indigo snake study.) After the eventful roundup, the three of us then went out to a local naturalist's home to get more information on a recent sighting, ate dinner at Pizza Hut (guess that was a favorite eatery for me back then), and parted ways on I-75 late in the evening as they headed north and I headed west toward Auburn.

This was one of the few times that I did have a federal vehicle, and wouldn't you know it: When I stopped to get a Coke in the tiny berg of Buena Vista, the daggone truck wouldn't start. I was sure enough stranded. Shortly after midnight, two local cops also stopped at the convenience store and noticed my plight. In neighborly South Georgia fashion, they woke up the local car mechanic, who came over and tried to start the truck—to no avail because the truck needed a new distributor.

By now, it was after 2:00 a.m. I had of course explained that I was an Auburn graduate student working on indigo snakes and that I had confiscated one at the Fitzgerald rattlesnake roundup. I told the policemen that I could sleep in my truck, but because it was a cold night, I needed someplace warm for the snake. After ruminating on this dilemma for a few moments, the cops came up with a solution: They would put me and my snake up in the nearby old abandoned jailhouse. (Apparently, a new one had been built somewhere else in the region.) The former jail was decrepit, but it had a heater and would be warm.

So I reclined on the bottom bunk in one of the cells, while the snake resided in his sack on the top bunk. I can't speak for the snake, but I got very little sleep. It was a small-town Saturday night, and I was serenaded into the wee hours by the rowdy locals, who partied just outside my window. The cops came back several times to check on me (and the revelers, I suppose) and would holler to see if I was okay.

First thing on that Sunday morning, I called Dr. Dan from the pay phone at the convenience store, and when I blurted out that I had spent the night in a jailhouse, he yelled for Ruth to come talk to me (as if it was something only a female could understand). Finally, I made the point that the co-op unit vehicle had broken down and I needed assistance. Needless to say, I was very glad to see Joe come to my rescue early that afternoon. Because we weren't likely to find a distributor anywhere in the Georgia boondocks, we had to tow the truck back to Auburn. In retrospect, I would guess that I am probably the only person who can say she spent a night in a jailhouse with an indigo snake. Sadly, after all my efforts on its behalf, the indigo snake later succumbed, most likely from the ill effects of gasoline. Although it is now illegal to introduce toxic substances into gopher tortoise burrows, some snake hunters in the past did indeed pour gasoline down a hose into a burrow to drive out any resident rattlesnakes. And this indigo snake was an unfortunate recipient of those noxious fumes.

Palmetto Flatwoods Blues

My time on the indigo snake study was winding down in the summer of 1980, but the snake gods had one more unusual adventure in store for me. Throughout the study, I had been more or less universally blessed by the helpfulness of Georgia folks in facilitating my access into potential habitat and specific sites where indigos had been observed. But there are inhospitable individuals everywhere, and I was about to encounter a particularly unfriendly one.

I was meeting with a Florida biologist who had seen an indigo snake in extreme southeastern Georgia three years previously. At a local convenience store, a helpful employee called around and let me talk to a contact who could supposedly get us into what was now a paper company VIP hunting camp. My field journal summed up that conversation: "Bad scene; man was hostile and rude; wouldn't let us come in; was concerned over that sighting made three years ago (ridiculous). I did some fast talking (awkward) and refrained from letting him talk to [the biologist]. First nasty person I've run into: the hell with him!"

The biologist filled out the habitat survey for me from memory, and after walking around in some nearby habitat that we could access, I thought that was the end of it. Moreover, looking for indigo snakes in this region of Georgia is truly challenging. Much of the southeastern corner of Georgia is a vast green sea of palmetto flatwoods. Saw palmetto is a common small palm with fan-like leaves that have sharply toothed stalks, making it difficult to walk through and certainly nearly impossible to find a rare snake because the thick vegetation basically obscures the ground. As I was to later learn in Florida, even finding the relatively large sandy aprons of gopher tortoise burrows is difficult.

But the southeastern Georgia chapter of my study was far from closed. Early one morning as I was getting dressed in my Opelika abode, the phone rang and a male voice announced that this was the Pentagon calling. I truly thought it was a prank by one of my Auburn friends; why would the Pentagon be calling me? Once I established that it really was a call from the Pentagon, I learned that an environmental impact study was being undertaken on what was then Naval Submarine Support Base Kings Bay. The navy had selected Kings Bay as its preferred East Coast site for the new Trident submarines, triggering a one-year impact study. And of course, documenting the presence of any federally endangered or threatened species was part of that study. Could I, asked the person on the other end of the line, come down to Kings Bay and conduct a search for indigo snakes? After I got over my initial shock, I'm sure that I explained that I was just a graduate student and would have to confer with my major professor/head of the Auburn Co-op Unit. It's important to note that it was August, a decidedly undesirable time to search for indigos because they move down from the sandhills into the edges of creeks and rivers.

Nevertheless, that early-morning phone call set the stage for an intense, grueling, and, not surprisingly, fruitless search effort. Dr. Dan and I called in the cavalry and put together a team of Auburn grad students and technicians. We literally searched until it hurt: It was miserably hot and humid. And especially in August, you can expect gnats, mosquitoes, yellow flies, horse flies, chiggers, ticks, and humongous banana spiders that spin their webs across trails. In this latter case, it's the web that's annoying because it wraps around your face and hair. As team leader, I was dubbed "Sarge" because I had to push these typically gung-ho field biologists to their limits. Donna even cried uncle, so to speak, and I tersely responded, as I dripped with sweat, "Do you think this is fun for me either?"

Despite all our suffering and searching, we didn't document any indigo snakes during our intensive week-long effort. We did find a number of other snake species, though, including an impressively large cottonmouth that camped out in front of us and gaped so that we could inspect, from a safe distance, the white of its mouth.

In the course of my career, I've often used an apt and succinct quote from a rocket scientist who was discussing the possibility of life on other planets: "Absence of evidence is not evidence of absence." There certainly could have been an indigo snake somewhere in those vast pine and palmetto flatwoods, but we had no evidence except for that nearby, three-year-old sighting by the Florida biologist. So I flew to Washington, D.C. (my birthplace), and was sequestered away in one of the Pentagon's auxiliary buildings in Alexandria, Virginia, to write up my official report. As is typical of federal documents, it was absurdly long, complicated, and unnecessarily anal-retentive in detail, especially since the bottom line was that we failed to document indigo snakes on the base.

Findings

So what scientific fruits resulted from my misadventures, copious amounts of sweat, and tick-and-chigger-bit body? Some 590 indigo snake sightings were reported by 173 persons; 511 were deemed valid by my classification system that considered professional and individual attributes of the interviewees. In a nutshell, most of the sightings were in a geographic province called the Tifton Upland and were concentrated along east-bank sand ridges of the Coastal Plain rivers and streams. Characterized by deep, sandy soils and sandhill vegetation (pine and dwarf oak), these ridges were created millions of years ago from windblown sand. Pretty amazing, when you think about it. Of course, I generated detailed maps of the sightings, some of which were published in the subsequent scientific articles. And I also documented the use of sandhills by indigo snakes during the winter months and the strong association of the snakes with gopher tortoise burrows.

My study didn't contribute only scientific findings to the literature on this imperiled species. It substantiated my hope and gut instinct that this career field, wildlife research, was the correct path for me. I was truly enamored with my master's research and honestly could have continued doing it for many more months. It fed my desires to work with critters, have adventures, and discover salient information that would assist in conservation efforts. I also learned that most scientific research, especially field research on wild animals, leaves the researcher with many more questions and therefore hungry for continued sleuthing.

Marriage on the Rocks

Although my indigo snake research was an overall labor of love (okay, maybe not the fruitless naval base search and Pentagon fiasco), it exacted a devastating toll on my personal life. I tried to keep my marriage going via letters and phone calls; alas, texts and emails might have helped, but they were years down the techy road. I also traveled back and forth to Albuquerque when I could: two long days of driving, about fifteen hours each. This was back when speed limits were fifty-five miles per hour on interstates, so the drive seemed to take forever.

The Christmas 1978 trip was stressful beyond my wildest nightmares. I encountered blinding nighttime snow in eastern New Mexico on the trip to Albuquerque and even worse conditions on the drive back east. On the latter leg, I crept along for hours at twenty miles per hour with the big trucks. My poor pups were wide-eyed in the back seat because the sound of the tires on ice and slush was so unsettling. When we limped into Amarillo and got one of the last motel rooms, Elsa, the corgi, uncharacteristically peed on the shag carpet because the ice outside just didn't seem like grass. That

I made it back to Auburn in early January was a miracle indeed.

But the weather conditions weren't the only stressor. It was obvious that the marriage was helplessly stranded on the New Mexico rocks. Mike had come to visit me and check out Alabama that previous summer, and though I had gotten the little Opelika house set up for us, he deemed the weather just too hot and humid. "My hair curls, and I look like a freak," he announced shortly after arriving. Mike had been working in various sales jobs for years, where appearance mattered, especially to him. We had both grown up in places with humidity, but I must confess that one gets used to the Southwest's lack of that cloying and annoying ambient moisture. At some point, as I realized that Mike wasn't going to relocate to small-town Alabama, I got creative, or perhaps desperate, and talked to deans at both Auburn University and the University of New Mexico (UNM) to see if I could do my class work at UNM and my research at Auburn. Alas, that was just too avant-garde. And now I was caught between a major professor who declared, "Damn women always quit," and a husband who threatened, "Come back or else."

Christmas 1978 was so strained that some of my friends asked if I wanted to come stay with them because Mike was angry and hurt. At the time, Mike was living in a rented cabin in the mountains east of Albuquerque. In a heartbreaking realization of our estranged relationship, we sat on the floor in front of the wood-burning stove as the poignant Barbara Streisand song "You Don't Bring Me Flowers Anymore" played on the radio. We hugged, as if to say good-bye, and tears streamed down our faces.

In a last-ditch effort to save the marriage of two people heading in very different directions, we went to a counselor. She summed it up so accurately and succinctly that it nearly took my breath away: "Mike, I see that you want to put Joan up on a shelf and take her down whenever you need her." "Joan, I see that you don't have the time for a husband with Mike's needs." Mike really did feel abandoned, and that was never my intention. My father had hit the nail on the head when he voiced his concerns that Mike might not be able to accept my career.

I surely did not plan, nor did I want, to get divorced. But in early 1979, Mike called me and said, "Did I tell you that I was getting married?" Say what? I think I would have remembered that. I knew he had an earlier interest in Mormonism, but that wasn't in play here. He had met a waitress with two children and thought this would provide the family environment he craved. (Sadly, it did not.)

My heart broke when the divorce papers arrived, and my good ole boy colleague Danny was the first to ask what was wrong. In a side note of humor, he hollered for his wife to come help deal with my upsetting situation and atypical tears. At the time, New Mexico seemed to make divorce disturbingly easy: Mike got the divorce and got remarried on the same day.

Back in Alabama, I still felt married and continued to wear my rings until, during a visit to my Maryland home, my brother Davie used cinema (the movie *An Unmarried Woman*, about infidelity and the end of a marriage) to convince me it was time to move on. The rings came off that night. Yet, as Mike realized that his impulsive move was not turning out as he planned, he would call and ask to reunite. But that bridge was irrevocably broken, and although my heart wanted to say yes, my head realized that he had made his bed; moreover, with two children involved, he needed to lie in it and not tear this incipient family apart. Over time, though, things did unravel. After many, many, years; six kids of his own plus three stepchildren; a near miss at some form of reconciliation when he and I were both single; and then his third marriage and my second one, Mike and I are now in a position to wish each other well and be friends again. And I think we both realize that we were just too young, and our life goals too dissimilar, to have ever made it to our old age together. But I thank him immensely for introducing me to New Mexico; that love affair only grew stronger with time and happily played a major role in my far-off future.

Fourth Transition: A Career in Florida

Long before I would reunite with my beloved adopted homeland of New Mexico, I still had *mi brillante carrera* (my brilliant career, totally tongue in cheek) in front of me. In October 1980, I turned thirty in Auburn, and my friends placed a unique birthday greeting in the Auburn paper, congratulating the "Norwegian snake handler" on this pivotal birthday. Around the same time, I was hired by the then Florida Game and Fresh Water Fish Commission to conduct research on gopher tortoises and possibly other harvested reptiles and amphibians. The commission had very few women biologists at the time, and the three-man interview team queried me carefully about my experience and philosophy; they weren't about to hire a female "bunny hugger."

Fortunately, Dr. Dan and one of my other mentors, Larry (a renowned wildlife biologist and gopher tortoise researcher who left this world too soon) had anticipated such questions and had therefore made sure that I had experienced both a quail and duck hunt. Somewhat to my amazement, I enjoyed these undertakings immensely, especially watching the hunting dogs at work. I still remember Gretchen, the yellow Lab, sputtering through the water before carefully retrieving the wood duck I had shot. And of course, my interview team recognized that I had looked down many a gopher tortoise burrow during the indigo snake study, so I definitely knew my way around a sandhill. The job would be mine.

8

Ancient Dunes, Black Holes, and Burrowing Turtles

A Dream Realized . . . and a Long-Term Study Begins

In late fall 1980, I relocated to Gainesville, Florida, to embark on a study of the Southeast's burrowing turtle, the gopher tortoise (*Gopherus polyphemus*). On December 5, I walked into the Florida Game and Fresh Water Fish Commission's Wildlife Research Lab; later, the agency would change its name to the Florida Fish and Wildlife Conservation Commission. Little did I know that this was to be my home away from home for a very, very long time. My quest had been fulfilled after a long and serpentine trail: I was finally a wildlife biologist. And the fun—and challenges—were just beginning.

Early on in my research, one of my many field mishaps included getting my commission-green, diminutive Chevy Luv truck hung up on a pine stump.

I had no recourse but to hike out to the nearest homestead. The Florida cracker who dispatched his son to rescue me wanted to know why I was out there in the piney woods on a Sunday anyway. I of course told him all about my gopher tortoise study. He shook his head and replied, "Well, you ain't got much to do if all you got to do is work on them ole gophers." I reckon he would be pretty shocked to learn that I conducted research on "them ole gophers" for over three decades.

Thanks to my indigo snake work in Georgia, I had more than a passing familiarity with what I jokingly referred to as the "black holes," the tortoise's deep refugia that wound down into the bowels of the earth. I knew how important these burrows were to a host of critters that needed to escape from heat, cold, dehydration, fires, and predators. I also knew that gopher tortoises were facing many of the same threats as indigo snakes, primarily habitat destruction and degradation. But at the time I started my studies, the gopher tortoise was still a harvested species. Many a family had survived the Great Depression, thanks to the "Hoover chicken" or "slow-walking chicken." The pursuit and consumption of this land turtle were part of the culture of the rural South, and in some areas, gopher meat was prized above steak. But could the tortoise survive sustained harvest, especially in the face of rampant development?

This was one of the many questions I was seeking answers to. And what other anthropogenic (human-related) factors were impacting this imperiled species? There were also a number of basic ecological questions: What are gophers doing in their day-to-day lives? Where and when are they doing it? How are they doing it? And most important of all, that brain twister, why are they doing it? Some excellent early studies by pioneering colleagues had set the stage for me, but, as I quickly learned, Florida's habitats and tortoise problems were often different from Georgia's. I must confess that I had to modify my paradigm of what was gopher tortoise habitat. It wasn't just the rolling longleaf pine–turkey oak sandhill that I had experienced in Georgia. It was the tall and twisted sand pine scrub that looked like a Chinese painting; it was vegetated and undulating white beach dunes; it was xeric (dry) hammock where the stately live oaks were festooned with Spanish moss; and it was a whole host of almost eye-dazzling-green, palmetto-dominated flatwoods and scrub.

At first, I just couldn't believe that any gopher tortoise would live in these nearly impenetrable habitats, and truth be known, they do prefer open-canopied, grassy, savanna-like habitat when it is available. But much of Florida is ancient dunes, the result of rising and falling sea levels over time, and those sandy soils support an incredible variety of vegetation. There are seemingly desert lands in Florida where you can trip over cacti, but the xeric nature of the habitat is due to *edaphic* (soil-related) factors rather than atmospheric aridity. For a neophyte wildlife biologist, answering salient questions about gopher tortoise status, distribution, harvest levels, and ecology in this vast and diverse state constituted a daunting challenge and frequently meant miles to go before I could sleep.

ANCIENT DUNES, BLACK HOLES, AND BURROWING TURTLES

So what was the best way to gather information on gopher tortoise status and to hone the prevailing knowledge of the tortoise's statewide distribution? Finding and counting the half-moon-shaped burrows with their accompanying aprons of sand was a way of gathering such data. Gopher tortoise burrows are more or less shaped like the rounded carapace (top shell) of the tortoise, and the aprons are excavated sand that these accomplished diggers bring to the surface. One famous researcher had actually ridden a motorcycle to do surveys of tortoise burrows. I figured out pretty quickly that I couldn't replicate that methodology without killing myself. And the data would be meaningful only if I could exactly follow his earlier survey paths. So I queried Lovett, one of Florida's premier wildlife biologists, who also happened to be a good buddy of my major professor. His answer seemed relatively simple: Learn everything I could via field surveys, soil and vegetation maps, interviews, and so on, so that I knew more than anyone else about what was going on with gopher tortoises throughout Florida. So that is what I did. But it was definitely no small task.

My experience with the indigo snake provided a good template, and once again, I sent out questionnaires and then followed up with interviews of biologists, wildlife officers, foresters, landowners, and tortoise hunters. Because there were no tortoise check-in stations (as there are with harvested deer), I had to go directly to the sources and accompany the hunters in the field. I also traveled many miles around the state to look at where tortoises existed—and where they did not. Finally, I did my best to determine the location and magnitude of impacts on tortoise populations: harvest, development, certain forestry practices, agriculture, and phosphate mining. One of our retired wildlife officers gave me a sobering historical perspective. After recounting tales of tortoises being exterminated by farmers and cattlemen, he commented, "You know, ma'am, the gopher hadn't had many friends till now." I also needed to assess the effects of habitat degradation, where canopy vegetation becomes too thick and reduces forage plants like broadleaf grasses and legumes. My common refrain would become, "No groceries, no gophers."

Fortunately, I wasn't totally alone in my efforts to understand and conserve these intriguing burrowing turtles. While I was out searching for indigo snakes in Georgia, a new organization was conceived and hatched in Florida: the Gopher Tortoise Council (GTC). Years later, when I presented an overview at the GTC's twentieth-fifth anniversary meeting, I noted that there were three noteworthy milestones in 1978: the first successful transatlantic balloon flight (by New Mexicans, no less), the first test-tube baby born, and the birth of the GTC. When I was hired in Florida, I became the state representative for this amazing group of dedicated scientists and laypersons and served in that capacity until my retirement. Over the years, this tenacious and respected conservation organization has promoted the protection and management of tortoises and the other upland species throughout the Southeast, has funded research efforts, and has been a major force in educating the public about tortoises. It was both a privilege and a pleasure to work with these folks, my esteemed colleagues/friends, on behalf of tortoise conservation.

Spending Time with Pullers

Because the harvest of this innocuous and iconic representative of Florida's sandy lands was quite controversial at the time, I initially focused my efforts on gathering as much information as I could about preferred hunting locations, sex ratios and size of harvested tortoises, and especially about harvest success and effort (how many burrows were targeted before a tortoise was captured). The hunters themselves were colloquially known as *pullers* or *hookers* (though I always preferred the former name). The term *puller* comes from the long rods, commonly made of PVC pipe, that were carefully threaded down the sinuous burrow passageway. In olden days, pullers might use a muscadine grapevine with a metal hook on it, or they might make a pulling device from the three-eighths-inch-long rod that forms the outer frame of bedding box springs (crazily inventive, I always thought).

Because that narrow rod wouldn't necessarily be comfortable to handle, a wooden handle was attached to one end. The other end would be bent in a ninety-degree angle so that it could slip under the tortoise's shell. An experienced puller was able to modify the

ANCIENT DUNES, BLACK HOLES, AND BURROWING TURTLES

metal part of the rod to accommodate the varying curvatures of the burrows. If all went well, the hook would catch on one of the rear marginal scutes (plates that run along the perimeter of the carapace), and the tortoise would be laboriously pulled to the surface. I quickly learned that pulling required patience, finesse, and strong arm muscles. Later, I determined that even the best pullers could only get one tortoise out of about every five to ten burrows attempted because of the tortoise's innate and incredible strength, burrow curves, roots, and many other factors.

One early significant hurdle in gathering knowledge about gopher tortoise harvest stemmed from cultural issues. How was I, a young white woman working for the state wildlife agency, going to infiltrate the world of Minorcan (residents of St. Augustine whose ancestors heralded from an island off Spain), peninsular black folks, and Panhandle white crackers that hunted tortoises? Certainly, my past experience with snake hunters would be helpful once I gained access, but I still needed to find cooperative tortoise hunters and convince them to let me tag along with them in the field. One of our wildlife officers helped me find my first candidate: Mose, a gentle, older black man. The officer had cited Mose in the past, but fortunately, Mose was a forgiving soul. He had no phone, so I had to go to his rural and quite rustic trailer ahead of time to set up a date to pull tortoises. He also had no air conditioning, so he liked to keep his trailer dark to somewhat shut out the intense Florida sun.

Like many Southerners, Mose appreciated a good whiskey, or perhaps any whiskey. After we had gotten to know one another, he would tell me to bring him coffee on the day before I wanted to him to pull. He would get "straight," and then after we finished sweating out in the field the next day, he would ask me to take him by the local liquor store. Because Mose knew I worked for the Game Commission, I wore my uniform and used my state truck. However, Mose understood that I couldn't very well park my official truck at the liquor store, so I would simply drive around the block while he was procuring his whiskey and securing it in a paper bag. Then I would drive him back to his dark retreat, which in some ways was reminiscent of a gopher burrow. I became fond of Mose, and when I would stop by to visit after I hadn't seen him for a few months, his face would break into a big grin and he would say, "I been seeing you in the spirits." I was never sure just what spirits he was talking about, but I took it as a compliment.

If Mose bordered on the angelic, another puller, Lennon, was the antithesis: a hell-raising young black man. I never would have gained access to Lennon without the assistance of a black secretary at the University of Florida School of Veterinary Medicine. I met her through veterinary school colleagues, and she kindly offered to take me down to her turf, the so-called quarters of Williston, a small town southwest of Gainesville. Lennon was her cousin, and even with her endorsement, he almost backed out on the day we were supposed to pull. I also owe Lennon's mama, who told him that the white girl sounded very nice on the phone and he should go pull with her.

Before the appointed day, Lennon's cousin, a matriarch of sorts and obvious pillar of the community, drove me all around the black section of town. At each corner where young black men congregated, she would introduce me and tell them, "This is my friend; don't hurt her." I had obtained permission from my supervisor, Tommy, to go in plain clothes; however, both Lennon and his cousin knew that I worked for the Game Commission. We had all decided that my driving around the quarters in a uniform and state truck would not be a good move.

After all the preliminaries, Lennon and I finally set out from town in his beat-up old car. He had a tendency to swerve a bit more than the rural road warranted, and I said a silent prayer. We headed to an old pasture with adjacent sandhill habitat. Once there, we climbed a fence, and Lennon proceeded to work his PVC rod down into tortoise burrows that showed signs of activity (tank-like tracks, shell scrapes). While I was scribbling notes about Lennon's technique, success, and other salient data, Lennon too was gathering information . . . about the life of a young white woman. I quickly realized that he hadn't spent a lot of time with white women, especially white professional women. After explaining that he generally went honky-tonking each night to local bars, he succinctly recounted his prowess and success as a local ladies' man in such graphic language that I silently gasped.

I was momentarily speechless, but not for long because Lennon really wanted to know about my social life: "So . . . what's you do in the evenings?" I tried to appear as nonchalant as possible as I explained that I had a long-distance boyfriend that I saw periodically, and I attended wildlife professional meetings where I interacted with colleagues. Otherwise, my fieldwork in the Florida heat was sufficiently taxing that I would go home, shower off the sweat and sand, and then sit down to read a good book. Lennon ceased his pulling for a moment, looked directly at me with a big grin, and said, "Ya know, I likes you even if you is dull."

As we moved from burrow to burrow, he achieved success in some cases, but more typically, the burrow was too long or crooked. The hot afternoon stretched into early evening before we had gathered enough gophers to satisfy Lennon. Pullers transport tortoises in burlap sacks, and we had several sacks with two or three gophers each. Lennon was obviously testing me, so he loaded me up like a pack mule, and off we went toward his car. But the fence, poorly constructed of both barbed wire and hog wire, still had to be negotiated. I climbed up on the fence with my heavy load and quickly realized that I would need to ask Lennon to hold my sack while I struggled with the half-collapsing wire. The inseam of my jeans caught on one of the barbs, and there I was with my crotch attached to the fence.

My embarrassment increased exponentially when Lennon, who had been "Mr. Cool" all afternoon, suddenly started trying to pull me off the fence and exclaimed, "What's folks gonna say if I takes you back all cuts up?" Now it was up to me to mollify the situation, as I reassured him that I would get off the fence without injury (though I wasn't at all sure that was true; I still have that earlier scar courtesy of barbed wire). Fortunately, I was able to safely extricate myself from the wire's grasp, and we loaded up the gophers and headed back to town. I don't know if the fence incident unnerved Lennon or whether it was just "Miller time," but he was swigging beer and swerving more than on the drive out from town.

We grabbed a bite to eat at Kentucky Fried Chicken, causing many heads to turn because we were, after all, a black man and a white woman who were covered with dirt. I had already measured and weighed the tortoises in the field, and I had to pretend that I didn't notice that Lennon was surreptitiously trying to make a sale when we got back to the quarters. Selling gophers was illegal at that time, but I was there to observe and learn, and I didn't have arrest authority anyway. Night was descending quickly now, and even with my protection mandate, it really wasn't safe for me to be out and about, so I thanked Lennon, bid him adios, and headed home to Gainesville.

But there was a rest of the story. Lennon's cousin had invited me to a gopher dinner later that week. After all, she now had fresh meat and felt I should experience the entire process from soup to nuts, or in this case, from field to table. So I returned to the Williston quarters and felt a wee bit like Steve Martin in *The Jerk*, welcomed in this African American family's home, but noticeably of a different color. The main dish of gopher stew was served and had a somewhat spicy bite to it, but it was indeed tasty. I recalled that famous author Marjorie Kinnan Rawlings had provided an early recipe for Minorcan gopher stew in her book *Cross Creek Cookery*: onions, peppers, and tomatoes were involved, but this stew seemed different. I also knew that most cultures that consumed gopher tortoises ate the forelimbs and hind limbs of their quarry, but some also ate the neck, tail, stomach, and unshelled eggs. Supposedly, the eggs contributed to superb cakes.

But I swear that one piece of meat floating in my savory sauce looked like a tortoise intromittent organ, a penis! Well, when in Rome and all that: No one was watching me like this was a joke, so I scooped up the piece and chewed . . . and chewed . . . and chewed some more. I also knew that gophers were considered medicinal or an aphrodisiac, as in "I'm feeling puny; go pull me some gophers." Perhaps Lennon had told his cousin of my paltry-appearing social life, but more likely, I had just received a desired piece of gopher meat.

As my study progressed, I was learning that gopher tortoises were more prevalent in the peninsula and were severely depleted in the Panhandle primarily due to past overharvest. I needed to gather information on current harvest way out west in Florida. Our fishery biologists in the Panhandle told me of possible pullers in their area, so I contacted Willard, who lived in the tiny town of

Holt, west of Crestview. Willard was an older, God-fearing, white man who had fed his kids and grandkids the meat of the "Hoover chicken." He agreed to take me on a gopher hunt on the vast Eglin Air Force Base, which covers parts of three counties and many square miles of sandhill habitat. The four-hour foray reinforced what I had already heard: Eglin especially had been hard hit by gopher hunters. In talking with natural resource managers at the base, I learned that even bombing range personnel, who had time between training missions, would go pull gophers and cook them. This explained why so-called closed areas, like bombing ranges, were as depleted as the areas open to the public.

During our hot, humid, meandering search, we found few burrows and even fewer that could be pulled. I must confess to feeling a twinge of sadness when the hunt produced two adult female tortoises. Depending on latitude, females take ten to twenty years to reach sexual maturity (closer to twenty in the Panhandle) and produce an average of only five to seven eggs, most of which are destroyed by predators like raccoons, foxes, skunks, and other varmints. These females were likely decades old, and they were the only ones left in a wide swath of habitat. My job was to gather information, not to judge those who still ate gophers, but the reality of this particular situation was more difficult to digest. After the hunt, Willard butchered the gophers and cooked them over a camp stove by the side of a little creek. He mixed the boiled meat with rice and butter and provided our picnic lunch. The remainder would be taken home for his grandbabies.

Willard and his wife were exceedingly kind and did their best to give me a historical perspective of gopher harvesting in their neck of the woods. In the past, there had been community gopher suppers, and I learned that some women bought their wedding dresses with money from selling gopher tortoises. At local stores, there used to be pens of gophers with their prices (a dime or quarter or dollar) based on size.

There was one awkward moment that evening when Willard brought out the Bible and asked me to read a passage about man's dominion over animals. I certainly have no problem reading the Bible, but I felt uncomfortable sitting there in my work uniform, reading about man using animals for his own purpose, no matter the scenario. It seemed a case of mixing church and state, since I was a state employee. So I tried to explain to Willard and his family that I believe God meant for us to exercise wise stewardship on behalf of our fellow creatures on this earth. Perhaps I made him think about wildlife conservation, though I doubt I changed his mind. I feel that he knew that the gopher tortoise had suffered at the hands of man, and like most of the pullers I interviewed, he knew that the harvest might one day be stopped.

I had another rather rotund, former Panhandle gopher puller tell me, "Ma'am, if you are what you eat, I reckon I'm a gopher." He went on to say, "But I don't eat 'em no more; now I protect 'em." He had realized that this slow-growing, slow-to-reproduce turtle couldn't sustain continued heavy harvest. So the once avid hunter had morphed into a staunch advocate and protector. Yet another former gopher puller recounted increasingly familiar tales of overharvest and habitat loss and reflected, "The day might come when it would be curious for my grandkids to see a gopher." My challenge, and that of other tortoise biologists, was to prevent that prediction from becoming a reality.

If I had a favorite gopher puller, it was undoubtedly Rufus ("Bubba," so dubbed like many a Southern man). Rufus was a Minorcan who had deep and long-standing ties to tapping the bounty of the land and sea. The Minorcans in St. Augustine, Florida, trace their history back to the eighteenth century, when a Scottish speculator set up an indigo plantation on Florida's eastern coast near New Smyrna Beach. After processing, the indigo plant was known as "blue gold" back then, and the dye was much in demand, especially in England, for its deep blue color. Most of the indentured laborers who were brought over to the colonies were from the Spanish isle of Minorca. And after nearly one thousand workers died from malnutrition, gangrene, malaria, and scurvy over a nine-year period, some three hundred survivors apparently abandoned this harsh hellhole and escaped north to British St. Augustine.

These folks were extraordinarily tough, and Rufus embodied that heritage. He was short in stature but stocky and extremely strong. I met him because he served as one of the Game Commission's contracted alligator trappers, meaning that he went out and captured nuisance gators and dispatched them. The trappers could then legally sell the meat and hides. Rufus had lost a digit to a gator, but even in his fifties, it didn't seem to affect either his gopher pulling or gator catching. His sense of humor was finely honed, and despite his tough exterior, he was kind and generous. Once he learned that I could use help catching gophers for my population study, he offered to assist me with his pulling rod and years of experience. So off we went to my sandhill study site west of St. Augustine.

That first year, he saved me many days of arduous bucket trapping (more on that in a moment) when he pulled a percentage of the targeted tortoises in a few hours. He also accompanied me on a professional visit to the Kennedy Space Center, where I was collaborating with biologists trying to understand gopher tortoise ecology in pine flatwoods and coastal scrub. His colorful jokes and witty comments left the space scientists speechless as we sweated in the hot, white sands while Rufus pulled gophers from deep down in the cool earth. He and I also became film stars of sorts in a BBC production that covered gopher tortoises, indigo snakes, and other imperiled wildlife. In the film, Rufus alluded to his live-off-the-land heritage, and I discussed the biology of the burrowing reptile that Rufus was capturing.

I interviewed many other gopher pullers, but one other who stood out from the rest was actually a state forestry employee who agreed to assist me later in my career with a tortoise disease study. As part of a collaborative undertaking, his boss allowed Larry to travel around with me for several days. Larry was a timber-cruising cracker from the Panhandle and had eaten gophers in the past. He still had considerable pulling skills, even though the tortoise had been fully protected for about a decade. When he got a gopher on the hook, he would turn his ball cap backward, glance up at me, and grin—and I swear he was a doppelgänger for John Belushi in *Animal House*. He was also one of the few guys that I let get away with calling me "Sug," short for Sugar.

My fascination with the lingo and localisms of rural Southerners was a common thread throughout both the indigo snake and gopher tortoise studies. For example, J. C., one of my snake hunters, called the large and magnificent piliated woodpecker "a Lord-God bird," as in "Lord, God, what a woodpecker." In the early days of my work in Florida, when I referred to a gopher tortoise as, well, a tortoise, one Florida cracker gave me that "it's obvious your education ain't done you no good" look and retorted, "Some folks think this here is a turtle, but it's not, it's a *gopher*."

Common names of critters are often confusing, which is why biologists resort to scientific names. But crackers have their own system of classification. I soon learned that this gopher was not to be confused with the mammalian pocket gopher, which was known locally as a *salamander* (pronounced *sallymander* and likely a corruption of the word *sandy-mounder* because the critter pushes up piles of sand). And this seldom-seen, underground mammal should not be confused with the amphibian of the same name, known locally as a *spring lizard*. So . . . to paraphrase comedian Jeff Foxworthy: If you know for a fact that a gopher is a shelled reptile, a salamander is a buck-toothed burrowing rodent, and a spring lizard is an often brightly colored amphibian, you just might be a redneck.

Distinctive Colleagues

Although my esteemed colleagues at the Gainesville Wildlife Research Lab were perhaps not quite as colorful as the gopher pullers or snake hunters I had encountered, they were certainly talented and dedicated field biologists who had a few endearing quirks of their own. Over my three decades in Florida, I worked with so many full-time biologists, technicians, and volunteers that to profile them all would require another book or at least a large appendix. I feel extremely fortunate to have shared my research and field adventures with such amazing wildlife conservationists. In fact, now that I'm retired, I truly miss the camaraderie and daily interactions that I had with my colleagues. In many ways, they were kindred spirits whose goal was to better understand and thereby conserve Florida's flora and fauna. We not only worked in the same lab but also coinhabited the proverbial trenches as we struggled to manage wildlife in the face of rampant development. Bouncing ideas off one another and discussing the ever-present biopolitics somehow made the struggle less onerous. We were comrades in the fight for life, *wild*life. Moreover, researchers are driven to find answers; it's what we do. For me, these colleagues contributed to my sense of family in this place where we worked: our lair, den, nest, and burrow.

Sometimes, these "family" members saved me—or in this case— my thumb. Chris was our panther biologist; his office was just down the hall from mine. At the time, I had an indigo snake, Zeus, that I used for educational purposes. One day, when Zeus grabbed my thumb after I had absentmindedly touched a thawing rat (his food) and then put my hand in his cage to change his water, Chris heard me cursing, came into my office, and carefully pried the tenacious jaws and sharp teeth from my hand. The herpetologists who would have readily assisted me weren't there that day, and everyone else in the hallway just stood there in shock as blood dripped all over the floor. I was very thankful for Chris's quiet and effective action. Jim, our furbearer researcher, quipped that Zeus probably would have stopped chewing when he reached my shoulder. (Trust me, Zeus wasn't that long but instead was built like a short, stout torpedo.) I had to go to the county health department to get a tetanus shot but suffered no lasting damage.

Two of my early and longtime colleagues/buddies were Paul and Ab, herpetologists extraordinaire. They went way back as friends, and while Paul was a full-time state employee, Ab was a university professor who was contracted to help with statistical problems and alligator research. Both were, and still are, stunningly sharp and resourceful field biologists. Paul had polio as a child and lost the use of one arm, but that hasn't deterred him from wrangling snakes, gators, crocs, or alligator snapping turtles. Paul even discovered a new species of frog in Florida's Panhandle, aptly named the bog frog. Paul has mentored (dare I say sometimes tormented?) me, especially when he was my supervisor for a while, and his red pen or font greatly improved my scientific writing. He really does know almost everything . . . and as he notes, if he doesn't know it, Ab does.

Ab is also so smart it's almost scary (in a good way) and pushed himself beyond human limits when afield; I know because I've been with him on some of these misadventures. There are so many

Paul and/or Ab stories, but one that stands out happened when all three of us went to Cape Sable at the tip of the Florida peninsula to check out the southernmost gopher tortoise population. Earlier in the day, I had kind of dropped my end of a large canoe (hey, they were heavy in the ole days), and poor Ab was experiencing reoccurring back issues. Now we were bouncing in the waves as our motorboat was making its way to this sandy cape.

Once on land, we quickly learned that *everything* on this faraway outpost pokes or bites you. The vegetation was prickly, and the mosquitoes were downright vicious. Fortunately, this was before Zika virus, but mosquitoes still carry myriad other pathogens. The tortoise burrows were unique and fascinating, with their rather large mounds of shelled sand. If the hot, incredibly humid, and buggy conditions onshore were less than inviting, we surely weren't going swimming because we observed shark fins offshore. Moreover, a huge thunderstorm was brewing offshore as well, and our visit to Cape Sable was cut short as we hightailed it out of there in ever-growing waves that slammed our small boat. I had only gotten a glimpse of this isolated tortoise population, one that likely was introduced by seafarers who carried the unfortunate shelled reptiles in their ships as a long-lasting, stored, food source.

My buddy and later bossman, Woody, was an alligator specialist but had knowledge of other crocodilians as well. I always teased Woody that he had been a diamond in the rough in his younger days; he was a rough-and-tumble soccer athlete and guys' guy until he met his delightful wife, Susan. Then he was polished to a fine sheen. On one memorable occasion when Paul, Woody, and I were doing crocodile surveys in the Keys, Woody had to run the airboat over a small grassy land patch to get to deeper water. Paul was on the other seat, and I was situated on the floor beneath him. I had little experience with the noisy airboats, and as Woody launched the craft over the rough patch, I reached back and grabbed onto Paul's pant legs lest I be flung out of the boat. Of course, I had to endure a wee bit of teasing about trying to "de-pants" Paul.

Over the years, a number of excellent technicians assisted me with my field studies. In some cases, these students or recent postgrad biologists went on to far surpass me in the depth of their knowledge on specific wildlife issues. Russ and Kathleen were early field assistants during the 1980s. Russ was an avid herper (he loved catching scaly critters), and one time he impulsively reached to grab a venomous coral snake that was crawling among the wax myrtle shrubs and low bush blueberry. I was equally quick in grabbing his arm and telling him that he was on my time and I wasn't about to fill out the reams of bureaucratic paperwork if he was bitten. I also almost bounced the poor lad (who is now a renowned professor) out of the back of my truck where he had been stashed when we had to give a ride to a timber worker whose huge log-laden truck got stuck because my little truck was partially blocking the sandy road.

Kathleen has always been an amazing "idea person"; she remains a good friend and staunch advocate for both wildlife and domestic species. She and I had our fair share of misadventures. My favorite Kathleen story involved a fateful day of setting tortoise traps at one of my far-flung study areas (more on those sites and undertakings later, but basically, we were digging holes and sinking five-gallon buckets at tortoise burrows). During the summer that Kathleen worked with me, she convinced two of her friends, brothers David and Bruce, to assist us in our trap setting on a sandhill site. Despite Kathleen's explanation that this was arduous work and that getting an early start was essential, the guys took forever to get ready and then drove equally slowly out to the site. Our late start meant that I had to crack the whip and keep things moving. On a humorous note, when we broke for a quick lunch (I never tarried during field lunches; there was too much work to do), poor David thought we were finished for the day. The look on his face when Kathleen announced (somewhat gleefully, I thought) that we were only half done was priceless.

Seared into my memory is an image of the guys out there with their Chinese-style broad-brimmed hats, looking like nineteenth-century railroad laborers working in the staggering heat and humidity. It was so bad that while I was digging around 1:00 p.m. (a nearly unbearable time of day in the sandhills during late spring/summer), I literally had to pour Gatorade down my throat at absurdly frequent intervals because I was sweating so profusely. Somehow, we all survived that brutal trap-setting day, and David and Bruce had an entirely new respect for what "girls" could do. I'm not sure how long it took them to forgive Kathleen for giving them a taste of hell.

During the late 1990s into the turn of the century, when I was conducting a tortoise disease study, I had two stellar technicians who have also remained my friends to this day. Both are radiant inside and out, and I was so incredibly fortunate to have them assist me in an extremely challenging study. Lori was aspiring to be a veterinarian at the time and is now a doctor with both a DVM and a PhD. She is a phenomenal vet with expertise in both clinical practice for canines and felines and in wildlife medicine. Her veterinary experience was a key asset as I learned to extract blood from tortoises (not turnips, but almost as difficult). Tough as nails in the field, with a "never-say-die," upbeat attitude, she helped launch the study to understand this tortoise disease and then greatly expanded the research after working with me.

We still laugh about the time I almost did us both in, so to speak. We were using a burrow scope (basically a camera on a long hose) to check occupancy of burrows and had employed a heavy tarp to diminish the light around the monitor. It was excruciatingly hot and humid in the sandhills that day, and it's truly a wonder we didn't succumb to heatstroke under the black canvas. Finally, we did have to throw in the towel when I noticed that Lori's face was scarily beet red. Since then, special boxes have been designed to reduce sunlight glare on monitors, but we were low-tech back in those days. Similar to when I was a firefighter and became sooty-faced, I always seemed to get more sand and soil on myself than any of the rest of my crew did. This prompted Lori to dub me "Joan Dirt," an appellation that never failed to delight her.

When Lori became immersed in vet school, one of my colleagues recommended Elina (whom he later married) to be my partner as a tortoise phlebotomist of sorts. Elina grew up in Finland, and despite the twenty-plus-year age difference, it's as if we were sisters from similar mothers in very different lands and times. She has worked on both bear and deer and now is Florida's premier deer research biologist. She is truly multitalented as a scientist, field researcher, writer, artist, and athlete. On a memorable and slightly unnerving occasion when a violent thunderstorm swept over the sandhills, I had to laugh as we took off for the truck. Her athleticism showed as she leapt gazelle-like through the botanical obstacle course of wiregrass, saw palmetto, small oaks, and leg-grabbing vines as I clumsily lumbered along behind. Luckily, the lightning spared both the swift and the slow-footed that day. Elina and I always had comparable eating habits (frequent, small meals) and used to healthily snack our way across north-central Florida as we drove between study sites. These tenacious technicians and others made the often rough field conditions more bearable and greatly enhanced the fun factor of demanding research projects.

Trapping Tortoises

In addition to determining the statewide status and distribution of gopher tortoises, a second major thrust of my research involved understanding population dynamics (how population structures change over time) and home range (movements and habitat use within an individual's chosen 'hood). To accomplish this herculean undertaking, I established two northern Florida study sites in the early 1980s. From 1982 to 1987, I captured gopher tortoises each spring on these sites—and then returned to both study sites during the 1990s and to one in 2009 to evaluate long-term population changes. I worked incredibly hard on these two sites. In fact, my bossman at the time, Tommy, used to tell me, "Young 'un, you done worked your butt off." To which I replied, "Tommy, you're not supposed to be looking." He would blush and grin, and we both knew that he meant no harm.

It was a fact that my uniform pants sagged at the end of what was often more than a month-long trapping season, and I was afield seven days a week. Why, you might ask, did I have to be out in the field every day? Capturing gopher tortoises, superb burrowers that can dig holes up to forty-plus-feet long and six- to twelve-plus-feet deep, requires diligence and tenacity. It involves drudgery that is unique even for a field biologist. A tried-and-true capture method is decidedly low-tech: Known as a bucket trap or Tarzan pitfall trap, it's basically a five-gallon bucket (smaller buckets or cans for juvenile tortoises) laboriously sunk directly in front of the burrow opening, covered with paper or foil, and then camouflaged with sand.

Of course, all traps were shaded with vegetation such as palmetto fronds or, more commonly, with tent-like shade covers made from plywood or roofing soffit (the latter being lighter and easier to carry). The rationale is that gopher tortoises will emerge from their burrows (whenever they please; it could take weeks) and hopefully tumble into the trap. If I had a dollar for every bucket trap that I set during my three-decade career, I would be rich. Depending on the substrate, installing the traps could be relatively straightforward or staggeringly arduous. Deep, sandy soils were easy to dig in, but the trap cover often collapsed before I could even walk away. Alternatively, soils with more organic matter and especially palmetto or tree roots were horrid. After the traps were set, they had to be monitored each day and sometimes twice a day.

In the fine art and science of capturing gopher tortoises, heat and rain (even though the buckets had holes drilled in them) were always a worry. In all my years of trapping, I never lost a tortoise to heat, and I lost only one tortoise to rain—a big female that fell in head first and drowned before I could get there. I had arisen early that rainy morning and had driven as fast as I could in a torrential rainstorm to the study site. Sadly, the paper cover had partially plugged the holes, and the rain was so heavy that the bucket couldn't drain properly even in deep sand. I was heartbroken because I always endeavored to keep my study animals as safe as humanly possible. And I was mortified because I had been accompanied that morning by a visiting wolf biologist.

Jane was incredibly understanding and consoling, telling me that she had lost wolves to tranquilizer darts. It comes with the territory, she reminded me; sometimes, despite all our efforts, we lose study animals to circumstances beyond our control. Biologists, of course, are accustomed to death. It is a part of life, but we do not want to kill the animals that we are studying and striving to conserve. Fortunately, I did not have to go through that experience twice, but I fretted during heavy thunderstorms, especially if the soil didn't drain well. I would pull buckets up if tropical storms were forecast or even if the soil seemed unusually soggy.

The latter was especially true on one of my study sites. Yes, gopher tortoises do generally inhabit drier ground, but much of Florida is a mosaic of uplands and wetlands. Moreover, the tropical climate can cause vagaries of weather where pine flatwoods, especially, can become flooded. During El Niño events, for example, I would receive calls from concerned citizens about tortoises in flooded burrows. Short of issuing each gopher a snorkel, there wasn't much I could do except reassure the caller that tortoises are resilient and adaptive (they will move to higher ground if they can). Interestingly, several studies have documented gopher tortoises using flooded burrows, especially in winter, possibly for thermoregulatory benefits. And I have observed tortoises pop their heads up from a temporarily flooded burrow, take a gulp of air, and then retreat into their subterranean refuge.

During my tortoise-trapping season, I added bucket traps on both study sites as previously inactive tortoise burrows showed active signs, but it was always the initial "bucket blitz" day that drove me and my makeshift team (typically colleagues or others we could conscript) to near exhaustion. Setting fifty to more than one hundred bucket traps in a single day is a stress-inducing undertaking, especially for me because I had to dig and place traps as well as keep track of what everyone else was doing. This was generally before GPS units, so I had to construct my own maps and then assign specific burrows (marked with both numbered ground flags and colorful pink or blue tree flagging) to each team member. Generally, four to six persons, including me, could "get 'er done." In some cases, I had to assure my volunteers (the nonbiologists) that I could indeed find the truck with its food and water. Vegetation thickness, tree roots, and soil consistency were always challenging variables. On one study area, the soil was deep and sandy as befits these ancient dunes; on the other site, we had to deal with a dark, bizarre soil that often seemed like cement and sucked the moisture out of our hands.

My Far-Flung Sandhill Study Site

My two primary tortoise study sites were as different as night and day. Roberts Ranch was a classic sandhill site, with deep, sandy soil; turkey oak; and longleaf pine. My study area, a small portion of this vast, privately owned ranch, was remote and off the beaten path. The closest town of any size was Palatka on the St. Johns River. In the early days, I had to cross a waterway named Oldtown Branch to reach my research site. This stream ebbed and flowed based on rain events. Although it obviously appeared somewhat wider during heavy rains, its crossability (can I get across safely in a truck?) was notoriously difficult to gauge. As a result, the stream was dubbed "It Ain't Deep Creek" by the locals, as in, "Why, that creek ain't deep; y'all go on through."

But the creek was deceptively deep at times, with large rocks and holes that seemed to pull vehicles into the depths of purgatory. On one occasion, soil scientists conducting surveys on the ranch just happened to be behind me in their large SUVs. I didn't have my usual truck (good thing: the diminutive Chevy Luv never would have made it across) and instead was driving an old surplus field truck that somehow had avoided being painted commission green. Known affectionately as "Big Red," this truck was beat-up and ugly, but I soon learned that it was one tough son of a gun.

I hesitated at the edge of It Ain't Deep Creek but knew that I would block the other vehicles if I didn't proceed. I also knew that the creek was up, and my colleague who was riding shotgun blanched as I stomped the gas pedal and the truck lurched forward, hit a hole, and tilted downward until water came up on the hood. "Gun it," my colleague yelled—and we bounced up and out of the hole to gain safe purchase on the other bank. We whooped and hollered and turned to see the first SCS (Soil Conservation Service) vehicle caught in the creek's tenacious grasp. I had many tortoise traps to check that day, so I needed to go on, leaving the dirt dudes to figure a way out of the creek.

On yet another high-water occasion at Roberts Ranch, I wisely left my Chevy Luv on the bank, carefully waded across with a backpack full of equipment, and checked my traps. The hours passed as I wound my way through the scrub oaks and wiregrass; reset traps that had been set off with no capture (a royal pain because the bucket was typically full of sand); and measured, weighed, and marked (using small holes drilled into the shell edge) those individuals that I had captured. Eventually, I headed back toward the truck on the far side of what was now a rushing river. To add to my "how the heck am I supposed to get across?" concerns, I was carrying several female tortoises in burlap sacks. Adult females are larger, on average, than males; these individuals were nine to eleven inches long and weighed about nine to ten pounds each. As part of my study, I determined whether individuals were juveniles, subabdults, or adults, and the adult females were taken back to the University of Florida to be radiographed (X-rayed, a safe and established technique for tortoises). The radiographs would tell me if the female had eggs and, if so, how many. This is known as the clutch size. Now I would have to get all of us across the torrent safely, which I did by working my way upstream until I could find a narrower, more suitable crossing that allowed me to hold on to branches and navigate over fallen logs. I was soaked, of course, but neither I nor the gophers ended up floating downstream in the drink.

Although sandhills can be harsh, unforgiving habitats, my Roberts Ranch study site had at least a few Edenesque characteristics, such as its creamy paw-paw flowers in the spring and tall, swaying, lavender blazing star blooms in the late summer. In one photo, I captured early-morning mist blanketing the smaller turkey oaks and swirling among the stately longleaf pines. And Roberts Ranch had at least one other claim to fame: as a VIP hunting reserve, it abounded with game species, especially white-tailed deer. But black bear also frequented the sandhills, flatwoods, and creek bottoms, and I would occasionally find burrows that had likely been dug up by bears. Had this been my other study site, I would have suspected nefarious humans or feral dogs, but the remoteness of my sandhill and the power required to dig up tortoise burrows led me to believe that bears were the culprits.

Very early one misty morn, as my technician Kathleen and I were sleepily bouncing along in the truck after successfully navigating It Ain't Deep Creek, our eyes popped wide open as a large black bear darted directly in front of the truck and lumbered up the hill above the creek. Without thinking, we both jumped out and ran after it. I'm not sure what we would have done if we had caught up with it, but fortunately, the bear wanted no part of two crazy female biologists who had suddenly come alive.

On another occasion, I was dispatched to Roberts Ranch to pick up a bear who had not been as lucky as the one Kathleen and I had seen in the back country. In this case, the large bruin had ventured onto Highway 100 at night and was struck down by a truck. The cop who awaited my arrival had already gathered several local men to help me lift the bear into the back of my truck. Highway mortality of large mammals such as deer, bears, and panthers is an ever-present concern for wildlife biologists and motorists alike in Florida (and sadly, elsewhere).

Suffering for Science

As fond as I became of my study sites and their resident gopher tortoise populations, there were times when the field conditions basically overwhelmed me. Misadventures were hardly uncommon (remember Murphy's law?), and there was indeed a learning curve to conducting intensive daily field studies in Florida's oppressive heat and humidity. One of the first things I learned was to invest in Gatorade, which coincidentally was a University of Florida–spawned creation. When I first started working at Roberts Ranch, I carried gallons of water with me, thinking that would suffice. On one particularly uncomfortable day, I consumed copious amounts of the now warmish water and went about my fieldwork. My stomach started gurgling, and as I was driving out of the study area on sandy, bumpy roads, I developed a serious bout of Montezuma's revenge and was forced to stop more times than I care to tell. Dealing with diarrhea in the field is no fun. I had learned a valuable lesson: you need to replenish your electrolytes via Gatorade. From then on, the blue or orange or red or yellow-green liquid was always in the cooler in the back of my field vehicle.

If serious dehydration or heat exhaustion wasn't enough of a concern, Roberts Ranch was renowned as a lightning site. Florida is apparently the lightning capital, and the high ridge above Oldtown

Branch was risky business during the summer months. The landowner had warned me on several occasions about avoiding those violent storms. I always had an eye on the sky to assess the location and color of clouds. Deep purple, bruised skies were often spectacular but potentially deadly. On one almost fateful day, I knew time was running out for me to check all my traps, but the welfare of my tortoises superseded even my worries about lightning. The storm swept in faster than I anticipated and caught me in a decidedly undesirable area among the now swaying longleaf pines. I had just captured two tortoises and didn't have time to work them up before the vicious electrical storm hit. All I could do was crouch down as low as possible, shielding the frightened tortoises under me, and pray as the lightning crackled above and around me. The wind was fierce, and the thunder was deafening; rain came down in sideways sheets. This is not the kind of storm that anyone should weather out in the open, but field biologists sometimes have no choice if their vehicles are far away. Fortunately, the lightning gods were benevolent that day, and I didn't sustain a hit. But I had learned another lesson and was more wary and careful thereafter.

Ironically, it was that same day, when I was nearly physically and emotionally spent, that my Chevy Luv decided to get irrevocably stuck in the mud on the far side of It Ain't Deep Creek. I hiked out five miles in the gloaming to the landowner's house and waited for a colleague to come rescue me. I had a bit of a wait considering it was one of the summer holidays and most folks were out recreating. We wisely left the truck there that night, and when my buddy Paul brought me back to extract the truck the next day, he admitted that he thought that I, the gal, would have just gotten one wheel into the mud and then declared the truck stuck. When he saw that the Luv was perpendicular to the road and deeply embedded in the viscous muck, he was impressed and commented, "I'll say one thing, Diemer, when you screw up, you really do it well."

Not surprisingly, many habitats in the subtropical Sunshine State harbor chiggers, ticks, mosquitoes, biting flies, no-see-ums, and other distasteful (to us—but not to all critters) entomological annoyances. The chiggers (red bugs) were especially prevalent and tenacious at my sandhill site; even odiferous sulfur powder sprinkled into my socks didn't seem to repel them. And I swear that the yellow flies seemed to wait until I was head-down, bottom-up, resetting a trap, before they delivered their painful bite.

There are many species of ticks that bedevil mammals, including humans, and then there are gopher tortoise ticks. This species is incredibly large when engorged, and seeing them attached to a tortoise's head is not for the faint of heart. During my 1995 follow-up survey, the Roberts Ranch site became renowned among my technicians and volunteers for what was simply dubbed "the tick burrow." Tortoise burrows are known to harbor many species of insects and arachnids (for example, black widow spiders), but this burrow's inhabitants were unlike any other I had seen in my career. No one, except me (as the principal investigator, proving that rank doesn't always have its privileges), was willing to reset the cover of brown paper over the bucket trap for this burrow. Why were my usually gung-ho comrades so reluctant? As one approached the tick burrow, things appeared deceptively normal. If the paper bucket cover and sand camouflage were intact over the pitfall trap, and one didn't tarry, that illusion of normalcy might prevail. However, if the bucket needed to be reset, or if one paused momentarily to reflect, the vibrations of that person's footsteps initiated an unnerving mass emigration.

As if sent forth by Hades himself from the depths of the underworld, dozens of ticks emerged from their dark recess. Without hesitation, they would commence crawling onto any stationary field boots; no matter that the boots weren't their real quarry. I always wondered if they gravitated to the sutures on the boots because of the ever-so-slight similarity to the sutures on a tortoise shell. On one occasion, I took our statistician, Steve, out to my study site to show him what I did on a daily basis. Upon seeing the ticks emanating from the burrow, he quickly pivoted and hightailed it back to the truck. Needless to say, we never did catch a tortoise in the infamous tick burrow. The shelled inhabitant had left old tracks but had apparently found a more hospitable refuge sans large numbers of bloodsuckers.

My Rough and Scruffy Study Site

If Roberts Ranch had at least some innate beauty, my other primary study site was considerably more an acquired taste. Lochloosa Wildlife Management Area was a pine plantation just north of the tiny village of Cross Creek. Famed author Marjorie Kinnan Rawlings's homestead was, and still is, intact and can be visited as a historical site. The nearby waterways and live oaks festooned with Spanish moss are undeniably scenic. But my study area was a mosaic of wet and dry soils, with tall, planted slash pines and scruffy clear-cut areas where the pines had been harvested. I grew to appreciate this cur dog of a site and spent considerably more time there because I conducted both a multiyear population dynamics study and a two-year home range study. The Lochloosa site was also much closer to Gainesville, making it easier to access on a regular basis.

Obviously, demanding tasks and misadventures abounded at my Lochloosa study site as well. During 1985 to 1987, I attached radio transmitters to and followed twenty-two tortoises of various sizes and both sexes. I tracked seven days a week during May and June and less frequently, usually twice a week, during other times of the year. That's a lot of hours in the field. In those days, I attached small radio transmitters to each tortoise's shell with dental acrylic powder; later in my career, I used specialized epoxy. Radio placement varied but was generally on the rear of the shell on males. For females, that placement could interfere with mating, so I affixed their radios near the front of the shell. On one occasion, my task of placing a radio on a squirming subadult gopher tortoise coincided with a visit by the head of my agency and other VIPs. Fortunately, they had a sense of humor because jokes were made about the "white powder" that I had put in a folded piece of paper and was carefully sprinkling on the radio.

When tracking in the field, I carried an H-shaped antenna and wore earphones to pick up the beeps emitted on a specific frequency for each tortoise. I truly believe my hearing was jeopardized from those sometimes ear-splitting beeps until I learned to wear lighter earphones that didn't cup over my ears. I located my radioed gophers by listening for the intensity of the beeps and by turning the antenna toward the sound. Factors such as rainfall and density of the pines affected my ability to discern the beeps and therefore impacted the ease or difficulty with which tortoises were located. Focusing on tracking a tortoise was often all-consuming and could exacerbate the potential for mishaps, such as tripping on logs or boot-grabbing vines and stepping into unseen burrows/holes or fire ant mounds.

In one case, I wasn't even tracking yet but tarried too long as I was checking a bucket trap and was distracted by finding a young coachwhip snake in the trap. I went to show this little gem to one of the local foresters who had stopped his truck to chat. Much to my embarrassment, I looked down to see that my lower pant legs were covered with fire ants. These introduced and problematic ants are aptly named: I had already sustained some painful bites and had to dart back into the woods and quickly remove my uniform pants to minimize the number of itching, burning, long-lasting blisters.

During the 1980s, including when I conducted my gopher tortoise home range study at Lochloosa, only portions of the pine plantation had been clear-cut. The majority of my study site was along a sandy road path with tall pines and oaks on either side. The roadside berms (banks) were favorite burrowing locations for my tortoises. The slope made constructing burrows a bit easier for them, and the apron of sand outside the burrow provided a porch of sorts, where a tortoise could bask or simply view the surrounding area. It's important to note that these burrowing turtles spend most of their lives inside their burrows; they come out to feed, breed, and lay eggs. Accordingly, when I tracked my gophers, a field notebook entry would typically read, for example: "#39 is still in burrow 18S" or "#133 has moved back to burrow 98S." At that time, the burrows were marked "N" or "S" for north or south of the main study area road.

What can appear to humans as mere holes in the ground are indeed essential refuges for gopher tortoises. Generally, gopher tortoises dig burrows that allow them to turn around at any point, so burrow size corresponds to tortoise length. As I was to learn from my studies, a constant game of what I called "musical burrows" (like the child's game of musical chairs) takes place. Gopher tortoises use multiple burrows over a given time period, and the number of burrows used is typically related to gender, relative maturity, and habitat quality. Adult males tend to use the most burrows. One of my Don Juan study tortoises, #133, used at least ten different burrows over the two-year study period as he courted his gal gophers. And disputes over favorite burrows do occur. Although it's universally one tortoise per burrow, I documented overnight, temporary co-occupancy by two adult males and could hear them slamming against one another inside a burrow that they both desired. Immature gophers will also sometimes temporarily cohabit in favorite burrows. And size does, to some degree, matter. In one case, an adult male blocked a juvenile tortoise from reentering its relatively small burrow. The bullying adult then enlarged the burrow and used it for several days before digging a new burrow nearby. The usurped wee one had to go find lodging elsewhere. Hey, it's every tortoise for itself out there in the piney woods.

The Saga of Dufus

Although each of the twenty-two radio-instrumented gopher tortoises taught me something about the species' movements, home range, habitat use, and behavior, I definitely learned the most from the diminutive wanderer that almost got me killed: #29, Dufus. Despite my early worries about Dufus when we struggled to get a signal via aircraft that horridly hot June day, this wee tortoise was living proof of the Tolkien quote: "Not all who wander are lost." I, of course, will never know why Dufus chose to emigrate during an extreme heat wave or how he navigated through a thickly forested, seemingly unfriendly, landscape to find his way to what must have appeared to be gopher nirvana: an open, grassy pasture. What I do know is that his story is fascinating and that he challenged my own tenacity to find and follow him.

I can confidently say "him" now, rather than "it," because later in the study when we were collecting measurement/weight data and replacing the radio in the laboratory (pronounced la-*bor*-a-tory during such procedures), this subadult became agitated and extruded his intromittent organ. Gopher tortoise penises have been likened to lily pads because this unexpectedly large and pinkish organ extends down and flares out in a surprising manner.

Dufus's gender was certainly one factor in his noteworthy emigration. Subadult and mature males do tend to roam (not atypical for many species, including humans). But there is a "rest of the story" surrounding Dufus's relatively long movement of 744 meters (nearly half a mile) in approximately four days. It was the longest recorded movement during my study, but hardly a record for gopher tortoises. It's important to stress that Dufus was not full grown and had relatively short legs when he made his now famous move.

I had actually first encountered Dufus three years before his emigration. At that time, he was simply #29, a juvenile that I initially captured in a can trap (I used smaller versions of buckets for juveniles) and marked in 1982. Determining a tortoise's age isn't necessarily easy, especially when they are adults, but I was able to see two rings on this juvenile's plastron (bottom shell). Those growth rings told me that #29 was likely about two years old. I recaptured #29 in late May 1985 in an adult-sized burrow along the study area road. After being fitted with a radio transmitter and released, #29 at first refused to go back into the burrow where he was captured and then moved into a nearby clear-cut the next day. My field journal reflected my confusion and surprise: "This is *weird*: strongest signal is *in* windrow in clearing." At that point, I couldn't find a burrow opening and wondered if the signal was being somewhat distorted by the windrow, a large pile of sticks, soil, and other forest-clearing debris. By the next day, #29 ceased being just another study tortoise and morphed into "What is this tortoise doing? What a Dufus!"

That pivotal name-change moment came on yet another scorching day when I was tracking as early as possible to avoid the worst of the soon-to-be one hundred–plus degree heat. Instead of getting a signal from ground level or below in a burrow (where one expects gopher tortoises to be), the beeps were still coming from even higher up in the windrow. My first heart-sinking thought was that some kind of varmint had grabbed Dufus while the tortoise was aboveground and had unceremoniously dragged the lifeless, half-consumed, body up into the sticks. Or perhaps the transmitter was now inside the varmint. Smaller hatchling tortoises have been referred to as "walking ravioli" because of their yellow-orange coloring, small size (about two inches long), and relatively soft shells. But Dufus was around five and was 153 millimeters (six inches) long, still fair game but becoming less vulnerable with age and time.

Yet predation was the only explanation my heat-addled brain could fathom at that moment. Carefully, I climbed up on the long, crisscrossed stack of sticks. I was totally shocked by what I saw. There, balanced like an Olympic gymnast on a beam, was a very much alive Dufus perched on a narrow log up inside the windrow. One misstep could have landed him upside down or sideways, caught in the intertwining pieces of wood. In my field journal, I had written: "Amazing; found #29 up in windrow, beneath branches. How did he climb up there? Didn't move body; could see some head movement; took several pictures."

At the time, I had no idea this tortoise was about to make a move of a lifetime, leaving the natal colony where he must have hatched and had lived during his early years. In retrospect, it's tempting to anthropomorphize and think like a human: that he was Sinbad the sailor trying to climb the proverbial mast to see what lay ahead. For five more days, this little high-wire tortoise migrated back and forth along the windrow. In one case, the signal came from a newly excavated burrow at the base of the windrow, and in another case, he was actually out on the ground in the sun.

Eight days after he had been radioed and released, Dufus simply disappeared. So began an intensive search for this missing tortoise. I walked for hours, trying to avoid the bring-you-to-your-knees heat of midday. Fortunately, some of my colleagues joined the search. When a radioed study animal can't be found, battery failure is always one explanation. But #29's signal had been strong and his radio was new. I recall having several state game trucks combing the paved country road next to my study area on one evening, slowly driven by one biologist while I or another stood in the back with the antenna. Alas, no signal pierced the humid but slightly cooler air of the descending night. Dufus had just seemed to vanish into thick air. With the other radioed tortoises, there were certainly days when the signal was weak or nonexistent, but I always located them within a few days.

Although aerial tracking is commonly used for large mammals, it's rarely needed for gopher tortoises. But now my determination to find this young tortoise led me to take to the skies on a blazing hot afternoon when no small plane should have been aloft (due to reduced lift and therefore the need for more thrust and distance to take off and land). Despite the close call, I was thrilled to have gotten a signal and to know that Dufus was somewhere in the vicinity. But the location was not on public land, and I couldn't gain entry without finding out who owned the property.

My next step, post–terrifying plane flight, was to go down to the county courthouse; remember, this was long before GPS, Google Earth, online property records, or other such easy-access avenues. My investigation revealed that the pasture belonged to a Mr. Wolfe. But I wasn't a game warden, and even in my uniform, I couldn't just march onto someone's private land and begin tracking without permission. While I was sleuthing at the courthouse and still checking open bucket traps on my study area, my colleagues Ab and Paul had driven down a narrow road next to the pasture and had determined that the signal was strongest near a pole barn.

Three days after the fateful flight where I truly thought I might die, I turned into the private driveway that led through the pasture and wound my way up to a sequestered, rustically elegant abode. A forty-something man with a limp and a twisted smile came out to greet me. I had found Mr. Wolfe, and after I introduced myself and explained the rather unusual story of Dufus, he agreed to let me track the wayward tortoise on his land.

Sure enough, I found Dufus's new burrow southwest of the pole barn. Obviously, Dufus was pleased with his new digs (literally) because he used that burrow for the next year. And it was equally obvious that this pasture was now his home because he used a total of four burrows in a relatively small area until the home range study ended in May 1987.

But finding and tracking Dufus wasn't the end of the story. In one sense, it was a beginning. Dufus's emigration got me thinking that maybe this pasture was some kind of a tortoise mecca and that I'd find other tortoises that I had failed to capture during some years of the Lochloosa population study. There were certainly a number of gopher tortoise burrows in the pasture. So, at the end of June 1985 and again in spring 1986, I set and monitored traps in Wolfe's Pasture which, by default, became another population study site.

However, no other marked tortoises from my adjoining Lochloosa site were ever captured in Wolfe's Pasture. In that sense, Dufus was unique and, by chance or by choice, had apparently joined a new tortoise colony. I reckon we all sometimes need new pastures, or perhaps the grass really is greener on the other side of the fence. Dufus couldn't have known what lay ahead when he got the urge to emigrate, so chance did play a role, but he had apparently chosen to settle down in that pasture.

As with almost all my studies, a fascinating human aspect was intertwined with the wildlife work. Mr. Wolfe proved to be quite a colorful character. I found out from other local landowners that although he was now more or less a pillar of the Cross Creek community, he had once been some sort of drug dealer and had accidentally driven his car off a nearby overpass. Badly injured, he rehabilitated himself in more ways than merely physical. He was undeniably an interesting if somewhat mysterious man, and in accordance with his name, he enjoyed mild flirting. He worked out at the same fitness center as I did, and he would come chat with me (and flirt) while I was toning my muscles on the machines. I would also see him hobbling along on a trail around his property with weights on his ankles, and if I was digging a tortoise trap, he would pause to chat about... well, life. He didn't seem to be married, but a nubile maid sometimes answered his door. And on a humorous occasion when he invited me and my then technician Kathleen to a neighborhood party at his house, I had to diplomatically deter his more intensive flirting directed toward Kathleen. ("Joan: *do something!*") I always kept a professional working relationship and am very grateful that he granted me access to both track and trap tortoises on his property.

Girl Gopher Power

In addition to solving the mystery of the whimsical, wandering Dufus, my tortoise home range and population studies were illuminating in so many other ways. An indelible theme that wound throughout my research was the behavioral plasticity of this species. These seemingly reptilian robots are indeed quite individual in their behavior, habitat use, burrow fidelity, seasonal movements, and myriad other aspects of their ecology. One might say they have "tortoise-ality" instead of personality.

Of course, there are averages and generalities, and biologists bank on those. But I personally found this individuality one of the most interesting aspects of studying gopher tortoises. Over decades of research, I and other biologists certainly saw individual tortoises move a kilometer (0.6 mile) or more. Typically, it was courting males who roved over relatively long distances, sometimes multiple miles. But I documented individual females, who are typically more sedentary, just take off and go.

I was surprised in 1992 when I stumbled upon a female tortoise who had been marked during the 1980s. No, she wasn't anywhere near where I expected her to be; instead, she was just strolling down a Jeep trail 1.3 kilometers (nearly a mile) west of the original study site. In another example during my 2009 follow-up study, a newly marked female left the ecotone (a transition between plant communities; in this case a mature pine plantation and a clear-cut) where the forage was pretty darned good, crossed a highway, and was recaptured two months later over a kilometer away in what certainly appeared to be grossly inferior habitat with minimal groceries.

Over and over, in other studies that I conducted as well, I saw these generally older, mature females seemingly retreating from the colony. In another radio-telemetry study that I conducted to determine the prevalence of a specific tortoise disease, I had to cut a trail back to a female that had dug a burrow away from the other tortoises in shrubs and vines so thick that Bigfoot could have hidden there. And after fearing that another radioed female at a state park had tumbled into a ravine, I scrambled down and through the swampy ground to find that she was sequestered away in a remote sandhill on the far side of the ravine.

In a nod to old Hollywood, I began calling these reclusive females "Greta Garbo" (as in "I want to be alone"). I have always wondered

ANCIENT DUNES, BLACK HOLES, AND BURROWING TURTLES

why certain females retreated from the main colony cluster and good vittles to settle (temporarily?) in less suitable digs. Moreover, these solitude seekers may move long distances to achieve whatever goal they have in mind, so to speak. I have discussed this behavior at length with my tortoise-researching colleagues (yes, the men too), and although it may sound sexist and anthropomorphic, I think these ole gals just get tired of being tormented by the relentless males, especially during the late-summer/early-fall scramble when males become obsessed with mating. I can understand the males needing to roam and find the most desirable females. But it's fascinating to me that some females, outside the spring nesting season when they must search for open, sunlit sites to lay eggs, just move off into the hinterland.

I was especially intrigued by the sister story of a radioed female Sonoran desert tortoise (my gophers' Southwestern relative) that, albeit with some assistance over railroads and other human-made obstacles, moved thirty-two kilometers (twenty miles) through the Arizona desert. Desert tortoises do move greater distances than gopher tortoises, but this was extreme even by the standards for that species. Alas, even we wildlife researchers cannot unlock all the mysteries.

I certainly don't want to give the impression that the majority of my female gopher tortoises were movers and shakers. Some were decidedly homebodies. During my home range study in the mid-1980s, I had a large radioed female, #90, who was so sedentary that a home range couldn't be computed for her. She used only two burrows (86 meters, or 282 feet, apart) during a nine-month period. Moreover, I recaptured her in 2009 in basically the same location, despite drastic habitat changes from forestry practices over the years.

In addition to sometimes moving away from the males, females are especially defensive of their burrows and will turn sideways at the entrance to block other gophers. This behavior is often most obvious during the spring, summer, and early fall, when males go courting. As noted earlier, male gophers are determined and tenacious, to say the least. It is difficult not to think that the females have just had enough. In one such instance, I came upon a burrow of one of my radioed females and heard a commotion just within the entrance. This big gal was literally shoving several males out of her burrow, much to my delight because I could easily capture them.

During this bucket-trapping season, one young male that I had recaptured numerous times along the road path, was one of the suitors being unceremoniously ejected. I couldn't resist shouting "You go, girl" to this fellow female. The way I see it (kiddingly, of course) is that this gopher tortoise was striking a blow, literally in this case, for females everywhere who've had their space usurped by males, especially in the bathroom and kitchen. I always found these behavioral observations quite compelling, a veritable *As the Tortoise Turns* or perhaps *Splendor in the Grass* soap opera out there.

Because gopher tortoises are fossorial, meaning that they dig burrows and live most of their lives underground, their day-to-day and seasonal natural history is even more difficult to document. Observing tortoises mating or laying eggs isn't necessarily common, but obviously with enough time spent afield, biologists do get to experience these intimate moments in tortoises' lives. On one memorable occasion in 1986, I was at my Lochloosa study site late in the day and had been accompanied by our newly hired computer guru, Jeff, who was curious about my research.

It was June 18, and we were tracking #75, one of the five radioed females. It was always exciting to see a study animal aboveground, and there was #75 out in a clear-cut (the same one that Dufus had used just prior to his emigration a year earlier). It quickly became obvious what she was doing: laying eggs in the sandy soil. I motioned to Jeff to stay back, and I knelt down some distance from where she was laboring under the intense subtropical sun. I didn't want to get too close and disturb her (and of course I didn't have my camera handy), but I could see her carefully swinging her back legs as she excavated a suitable nest cavity. One by one, I could see eggs slowly drop into the shallow hole: seven Ping-Pong balls, or at least that's how they appeared. Sweat dripped down my forehead into the curls tucked under my official ball cap, but I remained motionless. Much to my surprise, instead of heading directly back to her burrow in the woods near the edge of the clear-cut, she turned and walked clumsily toward me. Huffing and puffing from her extreme exertion, she kept coming until she was right in front of me. I stayed perfectly still as she (I swear!) looked up into my face as if to say, "That was a real pain."

Having not laid eggs or given birth, I couldn't empathize, but I surely could sympathize about such an undertaking in that suffocating heat. The scientist in me realized that maybe she was curious about this new obstacle out in her clear-cut so had deviated from her path home to check it out. But I also would like to think that perhaps this was a more mystical and spiritual interaction. Our techy guy was certainly impressed by the rare little drama that he was lucky enough to witness.

Alas, Mother Nature is not always benign, and #75's eggs were later uncovered and subsequently removed by an unidentified predator. As previously noted, tortoise eggs are tasty morsels for a host of varmints (raccoons, foxes, skunks, opossums, armadillos; the list goes on), and nest predation is extremely common. Truly, considering the high predation rates on nests and hatchlings, it is a wonder that any gopher tortoises make it to adulthood. It's a desert-like jungle out there in the Southeast's sandy lands, and getting large enough to avoid most predators requires lots of luck and knowing when to stay in your burrow.

Digging Out

My studies in a dynamic pine plantation gave me numerous insights regarding how gopher tortoises respond to various forestry practices. Early on in my research, a burning question was whether tortoises could dig out after an area was site-prepped by a large piece of machinery called a roller-chopper. As implied by the name, this piece of machinery chops up leftover debris and vegetation in preparation for planting the next crop of pine trees. A common type of chopper used in northern Florida at the time was a double in-tandem chopper (two large drum rollers) pulled by a tractor. In February 1982, my colleague Paul and I attached radio transmitters to three adult tortoises (two males and a female) that I had captured with a pulling rod. Trapping gophers isn't terribly fruitful in the winter. We released them into their respective burrows and allowed the chopper to do its work over the deep, sandy soil of this clear-cut. It was a bit disconcerting to see the obliterated burrow openings after the chopping; fortunately, the ground flags were still recognizable.

Over the ensuing weeks, I captured other nonradioed tortoises on the site in shallow, freshly dug burrows. That gave me renewed hope, and sure enough, all three of my radioed tortoises dug out during the eighth week post-chopping. I actually saw the female with her head down in an impression about fifteen feet from her original burrow. She stayed in that position for thirty minutes despite a very heavy April rain. The soil had collapsed in her original burrow, and there was little sign that a tortoise had exited. On a humorous personal note, I really had to pee while I was observing this tortoise from an agency SUV. I didn't want to disturb her, but the sound of the rain beating down on the roof of the vehicle was increasing my need to go. Finally, out of desperation, I peed as best I could into a Gatorade bottle. Obviously, I made sure to empty it at my earliest convenience so no one would mistake it for that not-so-delightful yellow-green beverage.

My study was a success: The tortoises were able to dig out following chopper treatment in deep, sandy soils, although their response could vary markedly in other soil types or with other site-preparation methods. The early reports of tortoise entombment from site prep likely stemmed from two factors: Evidence of emergence is difficult to discern and often fades with time unless the tortoise decides to reopen that particular burrow (which my female later did), and the disturbance often prompts some tortoises to relocate. I think tortoises sometimes tire of reopening their burrows and just figure that life might be easier somewhere else. I certainly saw evidence of that at my Lochloosa study site when the wandering local cattle would trample burrows. The tortoises often dug out and moved or started digging burrows under logs where their all-important refuges would be afforded greater protection.

Although I didn't have to worry about cows collapsing my home, I too needed to dig out of my burrow. I wasn't ready to become like some of my female tortoises that sequestered themselves away.

Already an outdoor girl on my uncle's apple farm in Michigan. Photo by my late father, Carroll Smith.

Protecting kittens from my inquisitive dog, Woofy, at our Maryland cottage. Photo by my late father, Carroll Smith.

Astride my horse, Flicka, on our Maryland homestead. Photo by my late father, Carroll Smith.

My wildland firefighting patch, complete with burnt edge. Photo by author.

Holding an indigo snake at the Wildlife Research Unit in Auburn, Alabama. Photo by the late Dan Speake, used with permission from his son.

Indigo snake crawling toward gopher tortoise burrow in Georgia. Photo by the late Dan Speake, used with permission from his son.

Burned site on St. George Island, Florida. Prescribed fire is the premier wildlife management tool for upland pine habitats throughout the Southeast. Photo by Cliff Leonard, used with permission.

Sandhill habitat with gopher tortoise burrow at Gold Head Branch State Park, Florida. Photo by author.

Positioning bucket trap directly in front of burrow to capture gopher tortoise. Courtesy of Florida Fish and Wildlife Conservation Commission.

Wire flap trap with captured gopher tortoise. Courtesy of Florida Fish and Wildlife Conservation Commission.

Measuring gopher tortoise with forestry tree calipers. Courtesy of Florida Fish and Wildlife Conservation Commission.

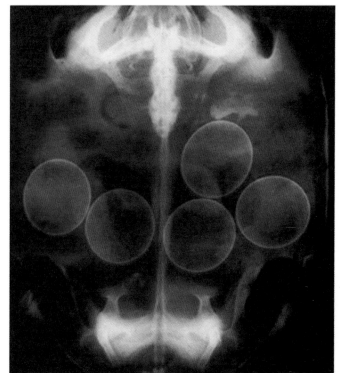

Radiograph of female gopher tortoise showing hard-shelled eggs. Courtesy of Florida Fish and Wildlife Conservation Commission.

Cessna aircraft with radio-tracking antenna attached to locate missing radio-instrumented tortoise. This flight was thankfully less eventful than the one to find Dufus. Courtesy of Florida Fish and Wildlife Conservation Commission.

Radio-tracking gopher tortoise to its burrow on my Lochloosa study site near Cross Creek, Florida. Courtesy of Florida Fish and Wildlife Conservation Commission.

Gopher tortoise captured on Egmont Key in Tampa Bay, Florida. Courtesy of Florida Fish and Wildlife Conservation Commission.

Adult gopher tortoise on my forestry site preparation study area, Florida. Courtesy of Florida Fish and Wildlife Conservation Commission.

Common bottom-up position of this tortoise biologist, with resident cat on Keewaden Island, Florida. I used the photo as an example of government pussyfooting around and having its head stuck in the sand. Photo by Deb Jansen, used with permission.

In the pit: captured gopher tortoise from burrow excavated by backhoe during relocation at Disney's Epcot, Florida. Courtesy of Florida Fish and Wildlife Conservation Commission.

Drawing blood from gopher tortoise held in the TRD (tortoise restraint device). Courtesy of Florida Fish and Wildlife Conservation Commission.

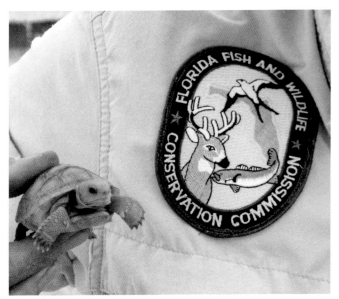

Hatchling gopher tortoise at my Wolfe's Pasture study site in Cross Creek, Florida. Courtesy of Florida Fish and Wildlife Conservation Commission.

Adult female gopher tortoise on my Plum Creek (formerly Lochloosa) study site, Florida. Photo by Cliff Leonard, used with permission

Juvenile gopher tortoise on my Plum Creek study site. Photo by Cliff Leonard, used with permission.

Inspecting marked adult gopher tortoise on my Plum Creek study site. Courtesy of Florida Fish and Wildlife Conservation Commission.

Releasing captured subadult gopher tortoise after collecting data. Courtesy of Florida Fish and Wildlife Conservation Commission.

Pine plantation habitat at my Plum Creek study site. Recent prescribed fire had fostered a lush ground cover and given this typically scruffy site a more eye-pleasing quality. Photo by author.

My Plum Creek study site is a mosaic of uplands and wetlands and supports many wildlife species. An anhinga (snakebird) is drying its wings and absorbing the sun's heat. Photo by author.

Holding a juvenile alligator captured in southern Florida. Courtesy of Florida Fish and Wildlife Conservation Commission.

Happily holding a rare Bolson tortoise at Mapini Biosphere Reserve, Durango, Mexico. Photo by the late Ray Ashton, used with permission from his wife.

Blue-footed booby in Galápagos, Isla Lobos. Photo by author.

Galápagos tortoise in pond, Isla Santa Cruz. Photo by author.

Keeping tranquilized bear's head above water in Cades Cove, Great Smoky Mountains National Park. Photo by Mark Stanley, used with permission.

Teachable moment for summer tourists in Cades Cove. Photo by Mark Stanley, used with permission.

Stormin' Norman destroying a Christmas toy, Alachua, Florida. Photo by author.

Merlin and Kotee interacting in my former Florida yard. Photo by author.

The otherworldly El Pinacate landscape, Sonora, Mexico. Photo by Mark Stanley, used with permission.

Pointing out a flowering cactus on a windy spring day in El Pinacate. Photo by Mark Stanley, used with permission.

Captain Dave with conch in Bimini, Bahamas. Photo by author.

Dave snowshoeing in Valles Caldera, New Mexico. Movies and TV shows (including the modern Western Longmire) have been filmed in this stunning volcanic depression. Photo by author.

Coyote in my mountain yard, Sandia Park, New Mexico. Photo by author.

Plump juvenile roadrunner in my New Mexico mountain yard. Adults are slender and long-legged and figure prominently in cartoons and folklore. Photo by author.

Mule deer buck making himself at home by drinking from the birdbath in my New Mexico mountain yard. He also carefully ate bird seed without breaking the ceramic dish. Photo by author.

Members of a bachelor herd of bighorn sheep lounging along road, Red River, New Mexico. My adopted state has diverse landscapes and abundant wildlife. Photo by author.

9

A New Life Partner and Many More Moons of Gopher Wrangling

My Quest for a Prince

While my professional life was churning along, sometimes at a seemingly breakneck pace, my personal life had stalled during the 1980s. I had dated, kissed some frogs that I thought were princes, even been engaged for a brief moment to a biologist; but disappointment over the "wrong guy" seemed to be the norm in those days. With no illusions of grandeur on my part, the hard reality is that it is often more difficult for educated, professional, assertive women to find a partner. And I was living in a college town where being thirty-something exacerbated the problem of finding a "mature" male. I had seen the same scenario with some of my professional women friends: Most guys were intimidated by their accomplishments. My friend Kathleen and I even tried a professional singles group. We met some interesting folks there, went on a few dates, heard the same lament from other highly educated women, but found no keepers.

Just about the time I was ready to give up on my search, one of my other women friends announced (out of the blue) that she had "the guy for me." I seemed to recall that she even gave me a list of this candidate's attributes. He was an accomplished sailor, scuba diver, spear fisherman, cave diver, and skydiver; he obviously had both an appreciation of the outdoors and a yen for adrenaline-inducing activities. And he even liked turtles. He had grown up in Michigan and had been an engineer in the Merchant Marine on the Great Lakes. He was now an air-conditioning, heating, and appliance repairman in Florida.

Although I was dating several guys I had met through the professional singles, I figured that this was a good time to play the field (I was, after all, in my late thirties now) and therefore accepted a dinner date with Dave. Hey, a girl has got to eat. We went to an upscale (for Gainesville) Chinese restaurant, and I learned more about his time on the boats. He also provided other details about his colorful past, when he experimented with hallucinogens in the 1970s. (I now tell him we could use those brain cells that he obliterated.)

The following morning, the phone rang before 7:00 a.m.; it was Dave and he just wanted me to know (at this hour of the morning?) that he wasn't like that anymore. He didn't do drugs. Okay . . . I mean, what do you say to that?

Shortly after my first date with Dave, my relationship with yet another Jim (I had dated three Jims more or less in a row: how weird is that?) heated up, and I wanted to see if it could progress; fortunately, it did not because this Jim was not one who could comfortably deal with a feisty, professional lady like me. During Christmas 1989, a memorable one because Florida actually experienced just enough ice and snow to make the roads a mess, my parents were enjoying the holidays at my abode, and I introduced them to several of the current candidates. My sweet mom liked everyone and rarely had any negative comments about anyone. But it was clear that Dave was her favorite. "Ooh . . . he's cute," she proclaimed.

I didn't see Dave for a while after the holidays, and life perked on. But air conditioners almost always need something in Florida. So one typically sultry spring day, Dave came to check out my AC. No, this isn't some kind of Freudian allusion; he really did work on my AC. I still laugh when I think about how he also took an interest in my refrigerator. As the relationship with the third Jim wound down to its inevitable conclusion during the summer of 1990, Dave resurfaced and asked if I wanted to go scalloping with him. He later admitted that he too was playing the field and had called several other girls to see which one was available. I was definitely ready for a new adventure, especially since I had never gathered scallops before (but I do like to eat them).

Off we went into the Gulf of Mexico to where the scallop beds were supposedly thick. I had borrowed a snorkel because the one

I had used back in New Mexico had more or less dry-rotted in Florida. The borrowed equipment had some kind of leak, allowing salt water into my mouth in greater amounts than desirable. The unrelenting sun coupled with the seawater entering my system made me weaker than usual, and after diving down and getting some scallops, I had trouble reentering the boat. Dave gallantly offered to assist, but our timing was off, and when I gave an extra kick with my fins and he grabbed me and pulled, I rapidly rose out of the water like a performing dolphin. Even worse, I basically bowled him over and landed spread-eagled on top of him. The fins that had proved so helpful in the water now made me even more ungainly in the boat, and I knocked him down again as we struggled to gain our feet.

He later said that he enjoyed that little fiasco, because I was clad in a bikini. As time wore on, the excess seawater caused my stomach to churn, and I eventually had to vomit, just a little, over the side of the boat. (Isn't that romantic?) My admin assistants back at the lab laughed for days after this misadventure. "Wow, you really do know how to impress a guy; you'll never see him again."

But Dave wasn't deterred, fortunately. Now our relationship was heating up, and it was accelerated, in terms of moving in together, by a truly bizarre event in August 1990. Gainesville, Florida, a typically quiet town in terms of serious crime, was dealing with multiple, blood-chilling, student murders by a serial killer. When a male student also became a victim after several horrific murders of females, my male colleagues who thought they weren't targets were suddenly reinforcing doors and pulling out their handguns and rifles. Palpable fear was pervasive because the killer remained on the loose. I would come home and check the closets before assuring that all doors and windows were secure. Poor Dave, as a repairman, was met at the door by gun-toting clients. Even my mother, who lived elsewhere in Florida but of course had heard about the student murders, deviated from her conservative nature when she asked, "I certainly hope Dave is staying with you." (Normally, she wouldn't have approved of "shacking up.")

So Dave and I moved in together in my tiny abode in the suburbs. We were hardly spring chickens (I was thirty-nine, and Dave was thirty-seven), and Dave had never been married. He had spent too many years as an engineer on the big boats (mega steel freighters), and although he probably had a girl in nearly every port, most women weren't willing to wait until he was done roaming Gitche Gumee and the other Great Lakes. As he told me, "I'm ripe for the picking." And I replied, "Well, you're almost overripe."

We got engaged in spring 1991 (yes, he did get down on one knee) and were married in a simple sunrise ceremony on New Smyrna Beach in July 1991. We had a brunch overlooking the ocean and receptions in Gainesville and Michigan for friends and family who couldn't make an oceanside sunrise wedding. Because Dave had been in the same Lake Superior storm that took down the freighter *Edmund Fitzgerald*, we had Gordon Lightfoot's famous ballad playing at our Florida reception. And although Dave and I epitomize the yin and the yang, total opposites, somehow the relationship has worked for more than twenty-seven years and is still going strong. The marriage gave me more than a new partner. For the second time in my life, I played the name-change game: From Smith to Diemer and finally to Berish.

Early in my quest, I made out a list of desired attributes in a mate, and Dave met a high percentage of the characteristics. Yet our approaches to many aspects of life are often different, sometimes diametrically opposed, and we regularly butt heads on issues. I often attribute that dichotomy to his upbringing. His large, boisterous, and wonderfully colorful clan marched to the beat of a different Michigan-based drum. That said, Dave is a true partner and an excellent grill master/cook; moreover, he was more than willing to help me with fieldwork, sometimes to humorous ends. I will always be grateful for his support of my tortoise research and demanding wildlife career.

One of my favorite Dave plus Joan plus gopher tortoise tales has served as an introduction to my professional presentations. I've used it as an example of the sometimes strange interactions between tortoises and humans in the Sunshine State. And although many Floridians think of gopher tortoises as inland inhabitants of sandhills, the reality is that tortoises occur in coastal scrub and even vegetated beach dunes not far from the waves.

In this case, we were over at Crescent Beach, south of St. Augustine on Florida's northeastern coast. It was early in our relationship, and we were strolling, hand in hand, on the beach. It seems there is no off duty for wildlife biologists in Florida, especially if one works on a high-profile, widely distributed, declining keystone species in a rapidly developing and high-tourist-volume state. My sun-infused reverie was abruptly interrupted when I spied a well-intentioned but obviously misinformed young man repeatedly throwing a gopher tortoise into the waves. When I marched into the gathering crowd and confronted him, he said that he thought it was a sea turtle. I introduced myself, explained my affiliation with the state wildlife agency, and gave him a succinct tutorial on the identification and habitat preference of Florida's upland turtle. It's a bit difficult to exert authority when one is wearing a bikini, but apparently my "Stand back, I'm a scientist" demeanor worked because he sheepishly relinquished the tortoise and allowed me to return the poor, sputtering creature to a nearby burrow in the adjacent dunes. Dave was impressed: Here was a gal that fit the Berish audacious MO.

Of course, our shoes were back at the van, likely being inundated by the advancing surf. The gopher tortoise's habitat in the dunes was home to the dreaded sand spurs, and trying to navigate in and out of the vegetation resulted in some considerable foot pain. Dave was beginning to understand that it was just another day in the life of a "gopher gal" biologist.

Population Snapshots in Time: Enlightening and Frustrating

The year after our marriage, I embarked on one of the two Lochloosa study area follow-ups. Dave was one of my volunteers, primarily on weekends during this demanding trap-a-thon. My original study area had been turned into a truly scruffy clear-cut: vast areas of leg-grabbing, low vegetation, dissected by long windrows. As the principal investigator on most of my studies, I called the shots and took on the responsibility for all aspects of planning and execution. That can be a bit difficult in the case of newlyweds. Fortunately, Dave was raised by a strong, assertive mom. She wrangled nine kids after Dave's dad died around the time of the birth of the ninth baby. So, taking orders from a female wasn't a big deal—but he also had sometimes unanticipated ideas of his own.

At the time, I was taking adult female tortoises back to the University of Florida to be radiographed so that I could determine the number of eggs. I was always careful about the welfare of my tortoises, but as I was checking traps one warm spring afternoon, Dave decided that one of the captured females being temporarily held in a burlap sack in the truck bed would do better in the deeper shade underneath the truck. Tortoises are Houdinis, master (or in this case, mistress) escape artists. When we both got back to the truck a short time later, the burlap sack was open and there was no tortoise. As I've previously noted, each capture requires a generally arduous trap installation followed by days of checking those traps. Having a tortoise escape before it is worked up (and in this case, X-rayed) was unacceptable. I was furious and told him not to come back until he had found that tortoise. By Jove, find her he did. He headed in the direction of the open sack and amazingly, she hadn't found her burrow yet.

After that, he may have paid more attention to my directions, but I doubt it. He always enjoyed foraging on blackberries in the field and seemed to delight in messing with the large ant mounds. He was an excellent bucket digger but was a bit overly fastidious (dare I mention OCD?) about getting clean after the dig. We had to bring extra jugs of water on days when Dave assisted in any aspect of fieldwork. It wasn't that he minded getting dirty; he just wanted to be able to get that dirt off before we headed back to civilization. And, of course, the poor guy was with "Joan Dirt." He also never understood the need to lock the truck when we were away from it. I often worked alone, and locking a state truck is always a good idea. And he was adamant about getting the AC turned on 2.4 seconds after entering a hot truck cab. During the 1995 follow-up at Roberts Ranch, the year of the tick burrow and just seemingly hordes of ticks in general, Dave had my colleagues in stitches when he kept pulling ticks off himself and finally looked up and exclaimed, "Are they falling from the sky?" For all his eccentricities, he was a super asset in the field and really loved those gophers, small and large. He would tease me by telling the juvenile tortoises: "Run, the witch is coming to get you!" But by then, of course, he knew better than to let one escape.

When I went back to my Lochloosa/Plum Creek study site (renamed for the timber company that now owned the land) in 2009, I had high hopes of reuniting with many of my marked tortoises from the 1980s and 1992 surveys. I was especially thrilled to recapture two male tortoises (in addition to the sedentary female, #90, previously mentioned) from the earlier home range study, still in their same approximate locations: #33 (first captured in 1982 as an adult: he was old) and #133. Interestingly, #33 had the smallest home range of the six males that I had tracked (0.23 hectare or about a half acre, and his home range was even smaller than that of some immature tortoises), and #133 was my Don Juan gopher that used the most burrows (ten) and had the largest home range (2.87 hectares or seven acres) of all twenty-two radioed tortoises. When handled, #33 was gentle and shy, while #133 was definitely not: he squirmed and urinated and defecated. This latter tortoise also had an extremely long gular, a frontal projection on the bottom shell used in male-to-male combat. In the tortoise world, I reckon he was quite the guy. In a scientific publication that presented my findings from the 2009 follow-up population survey, I noted that although I couldn't know their exact ages (remember, adults are difficult to age), these quite mature individuals present additional questions regarding the true longevity of a species estimated to live forty to sixty years.

Alas, despite capturing the three tortoises that I knew so well, the pragmatist in me realized that recapture rates over the long term aren't exactly scintillating. About a third of all the tortoises that I captured in 2009 were marked. That sounds encouraging until you look at the big picture: I recaptured only 8 percent of all 211 previously marked tortoises. There are many possible reasons for failing to recapture former study tortoises and then capturing new tortoises. The drill marks, for example, hold up well in adults but could be difficult to discern in tortoises marked as juveniles. Also, emigration (remember wee Dufus), immigration, recruitment (new babes), and mortality must be factored into understanding changes in any wildlife population.

Speaking of Dufus, I decided that I should also do some trapping in Wolfe's Pasture in 2009. Each day, my hopes of seeing the now mature male tortoise, #29, were dashed. To add to my frustration, my technician, Travis, was a jokester and couldn't resist hollering to me that he had indeed caught Dufus. It's a wonder I didn't fire him for teasing me so, but fortunately for him, I can take a joke. I'd like to think that Dufus is still somewhere in the north-of-Cross-Creek area, chasing those adorable females and generally getting into mischief.

Too Much Rain

The 2009 follow-up study was especially demanding in terms of the field conditions. The changes in my original study area as a result of silviculture (forestry) made it necessary to expand the survey area to try to find my tortoises. The site was now a checkerboard of intact pine plantation and clear-cuts. Moreover, the expanded site was also a mosaic of uplands and interspersed wetlands. To add to the inherent challenges, that merry month of May was absurdly and atypically rainy. I had to pull up bucket traps in some cases and replace them with a newly minted flap-trap that a colleague had designed. This rectangular wire trap is positioned over the burrow entrance and has a flap door on the bottom that is pushed up by an emerging tortoise.

These innovative traps seemed to work relatively well in the pine plantation of Lochloosa/Plum Creek but were not terribly successful out in Wolfe's Pasture. (So Dufus may have been there after all.) The relentless rains, so unexpected during the normally dry month of May, were reminiscent of some sort of prolonged tropical storm and made life difficult for tortoises and biologists alike. Portions of the study site that were dry in late April had standing or even flowing water three weeks later; it was totally crazy.

Thankfully, I lost no tortoises to drowning in bucket traps that spring. But I would wake up at 3:00 a.m. with the rain beating against my window pane, and I would fret. I actually ended up with acid reflux before that trapping season was over. The standing water reduced available burrowing habitat, and the tortoises moved to the highest ground, which was often old forestry windrow berms (embankments). The take-home message was clear: My tortoises really desired the better-drained soils and were not averse to climbing up and digging burrows in the top or upper sides of dirt-covered berms. Some individuals kept those elevated burrows as their alternate home during the high-water times. I even used the phrase, "If you berm it, they will come," with a nod again to Hollywood and the movie *Field of Dreams*.

Travis and I spent many hours soaked to the skin despite our commission-issued raincoats. Rain or shine, I was afield every day from mid-May to mid-June. When the field stress or worries about the tortoises' welfare during this aberrant weather led to my having heart palpitations (something I had experienced before during intense situations and especially super-humid times in the field), I figured maybe it was about time to wrap up my final tortoise trapping season. A part of me wanted to keep going, to try to find more marked tortoises and also more juvenile burrows. In the 1980s, the smaller burrows seemed to be lined up along the slightly sloped bank of the study area road. But now, finding those tiny burrows was nearly impossible with the abundant muscadine grapevines and other ground cover. The bottom line was that a larger tortoise metapopulation likely existed at my Lochloosa/Plum Creek area and that individuals moved in and out of the study site, responding to habitat changes and social or forage needs. I had to accept once again that we researchers often remain hungry for knowledge.

Blood and Nasal Fluid

During the 1990s and into the early twenty-first century, my tortoise research took on a new focus: upper respiratory tract disease (URTD). Tortoise die-offs were increasingly being reported, but little was known about the cause of these so-called mortality events. Obviously, this was yet another concern for those of us trying to conserve gopher tortoises. Researchers at the University of Florida's

(UF) School of Veterinary Medicine led the charge to understand all aspects of this recently discovered, but not necessarily new, disease: causes, clinical signs (symptoms), diagnostic tests, transmission, mortality, and so on.

My job was to gather information on the distribution of the disease and its effects on wild tortoise populations in Florida. This was no easy task, but I was blessed to have the amazing vet school team as collaborators and to have the cooperation and assistance of other biologists around the state. Thanks to my colleagues at UF, we had learned that the causal agent was a bacterium (not a virus) that lacks a cell wall, known as a *mycoplasma*. For animals, we use the term *clinical signs* because they can't articulate their symptoms; in this case, those signs included nasal discharge (what we might call a runny nose), ocular (eye) discharge, conjunctivitis (a reddened third eyelid), and palpebral edema (swollen eyelids). Basically, a tortoise with full-blown URTD looks, and likely feels, miserable: it has a snotty nose and moist eyes, and its eyelids are swollen and red. The disease is highly contagious and transmitted by close contact between tortoises. Those telltale signs may (or may not) appear one to two weeks post-exposure, but it takes six to eight weeks for an exposed individual to develop an immune response detectable by a specific blood test.

And this is where I fit into the picture. Initially, in the mid-1990s, I was traveling around the state and drawing blood from gopher tortoises. You might say I was a gopher phlebotomist (drawer of blood) or perhaps a vampire of sorts. I generally took about one-quarter to one cubic centimeter of blood from the brachial vein of the tortoise's front leg and transferred it to a special tube with lithium heparin to prevent clotting. I then centrifuged the blood and placed the resulting plasma (clear serum) into a storage tube. If I was going to be afield for several days, I had to carry a large liquid nitrogen container for the blood. That addition to my field gear often proved humorous when I was staying overnight at a motel and had to go out to the back of my truck in the evening. Fortunately, I was in uniform and using a state vehicle because otherwise I might have been mistaken for a witch with her milk-can-shaped cauldron as the misty nitrogen poured forth. The blood samples would eventually be processed at UF and would tell me whether tortoises in a given population or location had been exposed to the pathogen. Alas, it wouldn't tell me if the tortoises were currently infected, but it was a start in understanding the distribution of the disease.

Because I often had to work alone in remote areas, drawing blood from tortoises was rarely easy and was generally quite taxing: These individuals ranged from shy guys and gals with limbs and heads withdrawn, to leg-flapping, urinating, defecating, wild 'n' crazy ones. If necessity is truly the mother of invention, I needed an easier way to wrangle these gophers. Therefore, I conceived the tortoise restraint device (with the unintentionally humorous acronym TRD, pronounced "turd"), and one of our lab's wildlife student technicians built it for me. My colleague Paul promptly dubbed it my Rube Goldberg device, but it really worked, so much so that I and the builder (by then a federal wildlife refuge biologist) were later able to publish a paper on how to construct this portable device for restraining gopher tortoises during blood extraction. I designed it to fit on a pickup truck's tailgate. It was wooden platform with a block that served as a pedestal for the tortoise, supporting the plastron while preventing the tortoise from gaining traction with its feet. An adjustable wooden arm or clamp fit over the tortoise's carapace to effectively and humanely hold it in place. This practical device was far superior to two humans wrestling with a tortoise, and the original model served as a prototype for subsequent TRDs built and used by other researchers.

With TRD in hand (or rather on truck), I set off in the mid-1990s to gather gopher tortoise blood in the Sunshine State. In some cases, I or others trapped tortoises, but, as noted earlier, I also traveled with an experienced puller who could retrieve gophers from their burrows. The best and easiest capture technique was to simply encounter a tortoise away from its burrow. For example, in the vast Ocala National Forest, I would drive on linear open strips of land (known as gas pipeline rights-of-way) for tedious hours, hoping to see a tortoise out foraging in these openings that snaked their way through the thick sand pine stands. As crazy as it sounds, if I did spy one outside its burrow, I often had to dash like, well, the cartoon hare, to apprehend the little speedster before it could scoot down its burrow.

Gopher tortoises in the Ocala National Forest had a long history of being harvested by St. Augustine's Minorcans, so these individuals were quite wary. The antithesis was true on Egmont Key, an island and state park in the mouth of Tampa Bay accessible only by boat. As I raced around trying to capture the gophers that were nonchalantly strolling about, my buddy Dick, a renowned zoologist who knew this population, laughed and told me that these tortoises were so acclimated to humans that I didn't have to dash and grab. No one bothered the gophers in this large insular population, and as a result, they roamed about like Galápagos tortoises.

Although they weren't quite as tame as the Egmont Key tortoises, the beachside gophers in Smyrna Dunes Park on Florida's eastern coast also could be frequently viewed as they foraged outside their burrows. I totally wore myself out one very hot summer day as I would jump off a boardwalk or dune, grab a gopher, and hike out to where my truck was parked in the meager shade. I wouldn't work up a gopher tortoise out in the intense midday Florida sun, and if there was no decent shade, I rigged up an umbrella to shield the tortoise from the tortuous rays. After drawing blood from eight gophers that afternoon, I was so spent that I could barely make the two-hour drive home.

If tedium (as I searched for tortoises) or extreme fatigue (if I found too many) was a constant companion, there were also days that ranked way up there on the stress-o-meter or that were downright painful. At one of the Central Florida state parks, I lost a wee chunk of my forefinger's skin to a tortoise. That said, gopher tortoises are extremely gentle creatures, but hey, no one likes having blood drawn. In this rare situation, I told the park biologist, who was supposedly assisting me, that this large, parrot-beaked female tortoise was about to chomp down on my finger. "Nah," he said, "they don't bite." "She's going to bite me; please divert her head," I replied as I watched the syringe fill up with the bright red fluid from her foreleg. "Oh, they never bite," he said as she latched on to my knuckle. Never say never about any wild animal. I mumbled a few choice words and gave him a "you idiot" look.

But notched knuckles were nothing compared to the potential hazard of being on a "hot" bombing range at Eglin Air Force Base in the Panhandle. I was with my buddy Bruce, one of the base biologists, and we had trapped gopher tortoises in proximity to the target tanks (yes, they do burrow there) on the vast, minimally vegetated range. Bruce was in radio contact with the men who control the training missions, but the bottom line was that we were getting down to the wire on time. Not only did we have little time left to collect blood and return any captured individuals to their appropriate burrows, but we also had to get out of the area so that we didn't compromise a planned mission (that would have been very bad).

"We need to hurry, Joan," Bruce commented, calmly, I thought, under the circumstances. Somehow, my fingers flew as I worked the syringe on the last captured tortoise, glancing at my watch every few seconds it seemed. I jogged back to the burrow, making sure that the released tortoise went down deep into its protective "bunker." Bruce gunned the truck, and we spun out a bit as we raced over the sandy roads. It wasn't long after we exited the range that we heard the planes. Never a dull moment (well, sometimes) in the life of a gopher-bloodletting biologist.

A Tale of Four Study Sites

My URTD distribution study revealed that about one-third of the 386 tortoises sampled were seropositive, meaning that they had been exposed to the bacterium and had developed a detectable immune response. Although we now had some knowledge of where the disease was geographically, we still were in the dark about its effects on tortoise populations. For this reason, I embarked on an intensive, four-year study on four very different sites. One of those sites, Oldenburg Mitigation Park in west-central Florida, had experienced a significant die-off. My colleagues there found eighty-seven shells in June 1998. Three other sites in northern Florida were also chosen: Gold Head Branch State Park, Cecil Field Naval Air Station, and Big Shoals Wildlife Management Area. We knew that the disease was at Oldenburg, Gold Head, and Cecil Field; Big Shoals was to serve as my negative control site. To further understand what was going on, I would need to collect not only blood but also nasal fluid via a nasal lavage (flush or wash).

If you think that critters don't like having their blood drawn, try flushing out their nostrils. Needless to say, even with the TRD, two people are needed to accomplish this task, one to hold the tortoise's head and one to flush sterile saline into the nares (nostrils). The nasal sample could tell me if the tortoise actually had the mycoplasma in its nasal passages, meaning that it had URTD. Of course, each tortoise in the study would be evaluated for clinical signs. That said, another complicating factor was that clinical signs wax and wane, and some individuals remain asymptomatic (no signs of illness). This was definitely a complicated study, and as I mentioned earlier, I was truly fortunate to have ace technicians and volunteers to assist me. I captured tortoises on each of the four sites over four years (1998–2001) and also put radio transmitters on a subset to facilitate repeated sampling and to document mortality.

The four sites were quite different. Gold Head was primarily classic sandhill habitat but also had some desert-like scrub; Cecil Field was a mix of sandhill, flatwoods, and disturbed open land; Oldenburg was also primarily sandhill, but many tortoises were captured in a grassy power line right-of-way; and the Big Shoals tortoise area was an old field of planted pines tucked into a thick hardwood hammock. Each site brought different demands and misadventures (an understatement, to be sure).

Imagine, if you will, this scenario: You are strolling down a path through a verdant forest, enjoying the day, when an enormous metallic-green conveyance looms over you. Out jump two monstrously large beings that grab you and transport you back to the conveyance, whereupon they collect blood and other bodily fluids. Could this be the latest in a string of alien abductions? It's merely a day in the life of a gopher tortoise on one of my four URTD sites. And that is exactly how we started the study on Gold Head. My partner in bloodletting, Lori, and I were driving the main park road. After a wet El Niño winter, Florida now was dealing with drought and wildfires, with smoke permeating the humid air. As we came around a bend, we saw a stationary gopher tortoise near the forest edge. So we of course jumped out of the truck and grabbed her before she knew what was happening. She became #1, a radioed tortoise that I followed throughout the study.

On that fateful day when we first encountered her, she had a runny nose. Despite having URTD, #1 perked along and was sometimes seen roaming her part of the park road. I alerted the park rangers that she was a road walker, and I hoped that no one would run over her. She's also the female that I tracked through the branch (stream) that Gold Head is named for because I thought that perhaps she had somehow tumbled down into the ravine. Nope. She was "hiding" in a small sandhill patch on the other side of the stream.

This is why radio telemetry was essential for keeping up with these study tortoises. I spent many hours in the field at Gold Head, and in some ways, it was my favorite study site. Out in the hinterland, away from the main park road, cabins, and larger swimming lake, was a delightfully named tiny bowl of a lake: Devil's Wash Basin. Surrounded by scruffy and rough scrub vegetation, the lake's evocative name seemed to fit. On a few occasions, we would go have a quick bite there or at the shaded trailhead for the ravine. The roads in the backcountry were really sandy, and even with four-wheel drive, conditions could be dicey. On Mother's Day, Elina and I had a flat tire and got my truck stuck in the deep sands coming out of the scrub valley. One of the magnanimous park rangers, a mother herself, came with a bulldozer and extracted us before darkness fell. I was always so appreciative of the willingness of other outdoor professionals to go that extra mile to provide assistance.

If Gold Head was somewhat of a "golden child" near the quaint, lake-blessed town of Keystone Heights, then Cecil Field was the rough-and-tumble offspring. My study site was next to a minimally used naval air station runway and was at the end of a west Jacksonville road to nowhere. That dead end seemed to attract riffraff, and we'd been warned by a local cop to be on guard. Though we rarely saw persons back there during the day, there was certainly evidence of strange and unsavory activities. That reality was rudely brought home when Lori and I stumbled on what initially looked like a recumbent naked human or body in the foot-snatching tangle of blackberry brambles. When we got closer, we were relieved that it wasn't a dead human but were shocked to find that it was a full-sized, blow-up sex doll, with boobs and ruby-red lips. The briars had partially deflated it, but it still gave us the creeps to think about why it was there. As if the bizarre inflatable doll wasn't bad enough, we were walking a trail through the nearly impenetrable thick forest at one end of our study site—and there right in the middle of the trail to one of our bucket traps was a huge, fresh pile of human waste. Even my normally intrepid technicians weren't going to deal with that, so I had to go bury it.

We also found dead hogs on the site, and I wouldn't have been surprised to find signs of obeah (sorcery). I have inadvertently come upon such indications of Caribbean-based dark magic in extreme South Florida's disturbed and remnant woodlands. Cecil Field was also where, during an intensive trap-setting day, I had to temporarily cry uncle when I was suddenly beset with extreme heart palpitations. I was working away from the rest of the team and tried to lie down to get the palpitations to abate, to no avail except to render me fodder for chiggers and ticks. So, with my legs feeling like someone had filled my boots with cement, I stumbled back to the others and announced, "Houston, we have a problem." We broke for an early lunch, and after my palpitations had ceased, we resumed digging bucket traps into the sultry midafternoon.

The third URTD study site, Oldenburg, had been set aside as a mitigation park to help offset the loss of tortoises to development. Unfortunately, it appeared that someone had dumped sick tortoises into the protected park. The number of shells found in 1998 just didn't jibe with what one would expect that habitat to support. Because this site was farther from my home base and had agency biologists assigned to work there, I was really fortunate to have my colleagues Cyndi and Don share many of the radio-tracking duties.

On one rainy afternoon after a long atypically dry period, Don and I were checking locations of radioed tortoises. To our amazement, a three-legged female tortoise (she had somehow lost a rear leg) came hobbling down the power line path, stopping to drink and splash in the puddles. I had seen three-legged individuals before, mostly due to predation-related injuries. Although uncommon, these tortoises seem to get by, at least for some period of time. There was something both poignant and inspirational as we watched this little three-legged gal make her way along the sandy path, seeming to enjoy the rainstorm.

There was another noteworthy female at Oldenburg—a large, radioed tortoise that by all indications was very popular with the resident males. In fact, her popularity had apparently made her quite feisty, and we could actually more or less summon her by getting down and patting the sand on the burrow mound. We would hear her huffing and puffing her way up the burrow tunnel, and then she would stop just shy of the entrance when she saw it was humans and not her overzealous suitors. She was also more likely to give us a sneaky nip when we were sampling her, atypical behavior for most tortoises. Cyndi and I especially got a kick out of her "it's *my* burrow" antics, and she became one of our favorites.

Big Shoals, the fourth and final URTD study site, also yielded memorable moments and misadventures. This site was unusual in that much of the wildlife management area was not ideal tortoise habitat: It was a thick, shaded, hardwood forest. Fortunately for the resident gopher tortoises, there were human-made openings in this forest. One such opening was at the management area entrance and was patrolled by a frequently observed tortoise that my technicians and I dubbed "The Gatekeeper." But my study area was a mile or more down the winding and sometimes very muddy road, across a usually small stream, and out into an open field with various-sized planted slash pines. This field was a tortoise mecca, but my radioed tortoises did sometimes move into the adjoining hammock. (Guess everyone needs a change of venue now and then.)

Big Shoals bordered the Suwannee River (where there are actually rapids) and was used by cyclists, canoeists, hunters, and other outdoor enthusiasts; in fact, I had to help rescue hunters who had gotten stuck in the mud. At the time of my study, the state forestry folks had also contracted to have dead pine trees removed because of an infestation of pine bark beetles. We had to be very cautious on some of the road's twists and turns so that we didn't encounter a large logging truck at the wrong place. One of the drivers even backed over several burrows and just about ran us off the road a time or two. I did chew him out about backing over burrows, and that probably didn't endear us to him.

Speaking of terms of endearment, and I observed a memorable and humorous interaction at a quick-mart in White Springs after we had finished checking tortoise traps at Big Shoals. Field biologists spend much of their time in remote areas, and even though we carry food, water, and Gatorade in our vehicles, we sometimes have to reprovision at quick-marts in small towns. Most of these visits are relatively unnoteworthy, but in some cases, observations of the local folks add spice to the day. As we searched for snacks, the locals were catching up with one another. A mother and small child walked in and were noticed by an older man. Obviously, they knew one another, but it was the way that the man expressed his affection that jolted us for a moment. "Get your big-headed self over here!" he exclaimed. Granted, the youngster did have a somewhat large head for his tiny body, but would that description someday give him a complex? I doubt it, for he was still young enough to not have a well-developed ego. Nevertheless, the interaction was a source of amusement for Elina and me as we drove back to Gainesville.

But it wasn't stuck-in-the-mud hunters, road-hogging trucks, or colorful inhabitants that literally brought me to my knees; it was an unexpected turn in a downgraded hurricane. In September 1998, Hurricane Georges had hit the Mississippi coast, and it looked like it would meander north. In a fluke, it suddenly made an unexpected sharp right turn and headed east over northern Florida as a tropical depression. My team and I had just set bucket traps at Big Shoals the day before, thinking we were not in the hurricane's path. Now I was in high-stress mode, and despite the hazards of driving in a frog-strangling rainstorm, I had to check those traps. With me that crazy day was a renowned and often controversial tortoise biologist, my good buddy Ray.

Ray was an equal opportunity exasperator in that he spoke his mind and didn't really care who he offended in the process. He was a force of nature indeed and a good man to have riding shotgun during a scary storm. At the time, he needed some extra money, despite being somewhat of a jack-of-all-trades (consultant, activist, author, educator), and signed up to help me with my URTD study. Of course, that later became a problem when he also asked our Tallahassee office for a relocation permit while being employed by my agency (a conflict of interest).

For now, we were headed north on I-75 in a driving rain. The good news was that few others were certifiable enough to be out on the road. Even Ray, who was known to be loquacious, was uncharacteristically quiet as I struggled to keep the truck on the highway. I was relieved to get on the back roads in Big Shoals—that is, until I saw that the little creek was now a raging river. We had no choice but to leave the truck parked far enough away so the water couldn't reach it, pack up our gear, and hike to the study area.

And here's where things got dicey. Ray was a large man, so he stomped across and then turned and laughed as I slipped and went way down into the frothy water. Hey, it was pouring rain anyway, so getting wetter really didn't matter. We were both nervous about having a tortoise drown in a bucket, so we walked as fast as the conditions would allow. Thankfully, we had caught only one tortoise, and it was fine. We had no choice but to take the tortoise with us back to the lab, because working it up in those conditions wasn't tenable. And we lidded all the other buckets until we were sure about the path of this decidedly squirrelly storm. Somehow, we got all our work done and managed to get back to Gainesville safely, albeit soaked to the skin. I vowed never again to set traps if a tropical system was anywhere near the southeastern United States.

After four years of blood, snot, and copious amounts of sweat, we had some answers but still many more questions about the effects of URTD on tortoise populations. Even four years is only a snapshot in the life of a species that can live forty to sixty years. My team and I had captured and sampled a total of 205 tortoises and had put radio transmitters on 68 individuals. The tale of four study sites revealed a spectrum of serological (blood) results, clinical signs, and potential effects related to the presence of the mycoplasma.

For researchers, publishing is paramount, and my colleagues and I were able to publish a detailed account in the *Journal of Wildlife Diseases*. At one end of the spectrum was Big Shoals, where no tortoises were seropositive or showed nasal discharge. And although some radios failed over time, causing us to lose contact with individuals, we didn't document mortality there.

At Cecil Field, the percentage of seropositive tortoises remained low, clinical signs were not uncommon, and mortality was relatively high: six of sixteen radioed tortoises were found dead. An unusual finding was that the species of mycoplasma was different at Cecil

Field; unfortunately, we couldn't say unequivocally that this bacterium caused the deaths.

At Oldenburg, which had experienced an extreme die-off prior to initiating the study, annual percentages of seropositive tortoises fluctuated greatly (from 76 percent to 8 percent), clinical signs were very common, and five tortoises died. URTD was certainly present here, and it's possible that our study might have occurred as the population was recovering from an acute epizootic (a disease event in animals, analogous to an epidemic in humans).

Finally, the percentage of seropositive tortoises at Gold Head fluctuated over time but remained the highest of the four sites. We found no dead radioed tortoises, but some radioed individuals went missing. Nasal discharges were common, and the presence of the typical mycoplasma was indicated by nasal lavage. We felt that URTD was chronic in this population.

One frustrating result was that none of the eleven dead tortoises (at Cecil Field and Oldenburg) were positive for the presence of mycoplasma in their nasal passages. That particular finding ties into the mantra "absence of evidence is not evidence of absence" and may be due to the difficulty of getting a good, usable nasal sample. All eleven dead tortoises had been found aboveground, except for one that we exhumed from about six feet down in a burrow. One school of thought is that these typically burrowing reptiles come out because their thermoregulation ability is impaired, so they might try to bask under adverse conditions. Or these ill animals may just have been moving between burrows and were too sick to make it home. Although we had hoped to view the effects of URTD in a more black-and-white manner, even the shades of gray that we found were valuable and provided a baseline for further monitoring.

Singin' the Tortoise Relocation Blues

If the tortoise research fed my soul by allowing me to answer questions and satiate my innate curiosity, then the other "R" word, *relocation*, truly seared my soul over my thirty-plus-year career. Initially, the challenge associated with relocating tortoises out of harm's way (development) was only a sporadic thread in my professional life. But as time wore on, relocation sometimes completely dominated my day-to-day work as my agency's policies and protocols evolved and became, in some cases, so detailed as to be anal-retentive. And not surprisingly, the discovery of upper respiratory tract disease further complicated the already difficult undertaking of moving turtles that burrow deep into the earth.

At first, I was able to tie early relocation efforts into my research: Where were tortoises being most impacted by development, and how did they respond to being uprooted and relocated? One of my initial experiences with relocation came when Epcot was being constructed on Disney World property. The access road into Epcot cut through scrubby flatwoods inhabited by gopher tortoises. Neither Disney nor DOT (Department of Transportation) wanted to bulldoze and pave over these charismatic and increasingly high-profile critters. That would be akin to killing ET! So we employed backhoes to dig deep down and rescue these subterranean reptiles. This was my first time down in the trenches, literally, of a backhoed burrow (though I had of course hand-dug burrows to capture indigo snakes).

Because flatwoods do flood on occasion, some burrow bottoms were wet, so we had to laboriously extract the frightened tortoise from under water. In other parts of this highly variable site, we dug so deep that I thought we must be nearing Shanghai and found the tortoise still in dry sand. We were able to rescue and relocate thirteen tortoises on Disney property, but despite all the noble intentions, the bottom line is that development reduces the overall acreage of available wildlife habitat. Yes, we all need homes, and humans seem to need more than their fair share of space for work and play. But after many years of watching the habitat base being eroded, I conceived the slogan "Save space for wildlife."

I've always maintained that tortoise relocation is decidedly not a black or white issue; instead, it wears many shades of gray. If the Disney development was at least an example of tortoise relocation success, other early attempts to move tortoises seemed ill-fated from the get-go and emphasized additional biological and practical considerations. In one case, I was serving as my agency's link on a tortoise relocation by a Jacksonville utility company. The relocation should have been fairly straightforward. The company biologists would bucket-trap the tortoises and move them a short distance within the property boundaries. Unfortunately, the tortoise habitat was bulldozed the day after the bucket traps were set; that is not the way it's supposed to work. Thus, this unsuccessful endeavor strongly pointed out the need for proper coordination between company biologists (or whoever is conducting the relocation) and the construction personnel.

In an even more frustrating fiasco, I was pulled into an already problematic relocation in South Florida. A golf course was being constructed in Ft. Lauderdale, and tortoises were being picked up opportunistically by construction personnel and given to a wildlife officer who placed them in a nearby state park. It's important to note that such undertakings were before tortoise relocation really came onto the scientific community's radar. That said, dumping tortoises onto protected lands isn't the solution to the displaced tortoise problem. Habitat suitability and potential overpopulation of protected lands must be factored into a decision to move tortoises.

As this South Florida development heated up on the tortoise-concern meter, I was called in to assist. At the time, there seemed to be a need up in North Florida to stock tortoises on a federally owned, sparsely populated island. This could be a win-win situation, or so I thought. My buddy Ab and I labored for days under the intense subtropical summer sun to capture fifteen additional tortoises from the proposed development site. One of the tortoises, whom we named TR (for Teddy Roosevelt because ole Tall Tale Teddy had supposedly found a record-size tortoise), resided in the motel bathtub (sans water) at times because he needed a wee bit of breathing room. Despite all our sweat and arduous bucket-trapping, the so-called restocking was abruptly deep-sixed, so to speak. At the eleventh hour, concerns about long-distance relocations and potential mixing of gene pools arose from within that federal agency, and I was basically told to "stand down" by my agency.

That left me in the uncomfortable position of having to find a home for these wayward tortoises on the heavily urbanized southeastern Florida coast. I finally identified a sand ridge about fifty miles up the coast, and I was told to release the displaced tortoises without further ado. I sadly watched these individuals wander slowly away, looking right and left for familiar landmarks that didn't exist. I later learned that our desperately targeted relocation site was slated for eventual development. Another bitter pill to swallow and a lesson learned: Always procure and lock in a biologically suitable and secure site prior to any relocation.

As illustrated, the early days of gopher tortoise relocation were incredibly frustrating and generally unsatisfying from a tortoise conservation standpoint. Even before biologists got involved, everyone from John Q. Citizen spying a roadside tortoise; to gopher hunters who wanted a private stash on their "back forty"; to well-intentioned park rangers, wildlife officers, and other conservation-oriented folks had engaged in willy-nilly tortoise relocations. In one noteworthy case, foresters with a large North Florida paper company apparently relocated five hundred to six hundred tortoises over the years because "them ole gophers" were raiding the company's wildlife food plots.

Racing Tortoises?

To add to the sheer craziness, tortoises were also being gathered in large numbers, raced, and released wherever it was convenient for the race organizers. Although it would be difficult to pinpoint an exact date when Floridians began pitting these reptilian speed demons against one another, organized races to supposedly benefit charities had been in existence for several decades when I first came on with Florida Game and Fish in 1980. Of course, I also got pulled into that tar pit, which entailed trying to figure out where the corrals of race gophers had originated and where they might go after "running" their races.

Moreover, tortoises gathered up and used for racing purposes further highlighted concerns about population disruption, gene-pool mixing, possible disease and parasite transmission, and humane treatment. For example, it's likely that the initial discovery of upper respiratory tract disease on Sanibel Island in southwestern Florida stemmed from the release of large numbers of former race tortoises that had been gathered from as far away as Georgia and had been closely confined for lengthy periods.

At one point in the heyday of gopher races, I answered my office phone to find one of my agency's regional managers on the other line. "Hey, sweet thing," he drawled in a deep Southern accent. "You remember that pretty little dress you wore to the last commission meeting?" I held the phone away from my ear for a moment and then replied, "Actually, I was wearing a business suit." (I had gone to a lot of trouble and expense to find just the right combination of professional attire with a somewhat feminine flair.) "Well, no matter, could you put that on again and come on over here and sweet talk these ole boys about this gopher racing issue?"

I counted to three and replied that I would take the matter under consideration with my supervisor. Later, I learned that this high-up in the agency had been reprimanded by an even higher-up, the top dog at the time, who had heard that one of his regional managers had called me "sweet thing." I knew that this regional manager was stuck in another era and that he meant no harm, but I did not dress up and go pull his hind end out of the fire set by the good ole boys who wanted to continue racing gophers despite the uproar by conservationists.

Eventually, our agency told the race organizers to come up with an alternative to using live gophers in races. And, amazingly, they did. In one case, these creative country boys came up with wooden gophers that operated on a pulley, and in another case, they conceived and constructed bionic gophers (like remote-controlled race cars) that looked like turtles but would pop a wheelie and roar off down the track. The times, they were a' changing.

Don't Shoot Me, Bro!

By 1983, it became obvious that we biologists needed more information on tortoise relocation success: How well did these transplanted individuals settle into their new digs? Before I was hired by Game and Fish, my UF zoologist buddy Dick had been involved with a relocation of tortoises from a sandhill slated for mining by Dupont to a reclaimed mining site on the Camp Blanding Military

Reservation in north-central Florida. Now, five years later, Dick and I were going to attempt to find out if any of the relocated tortoises were still on that reclaimed site. I should note that reclaimed mining sites look almost nothing like the natural habitat that they replace. They tend to resemble vast rolling pasturelands or old fields, and the soil is strangely colored and textured. Dick and I set bucket traps in May, and the day after we set the buckets, I returned alone to a find a scene that seemed to emanate from an old war movie.

The quietude of the previous day was gone, and I was totally shocked to see that my study site was crawling with dozens of gun-toting soldiers. At the northern edge of the tortoise capture area, large guns were being fired into the distance. For a moment, I froze. I quickly realized that I couldn't enter the study site at my usual point, so I opted for another entry site farther east on the paved road. I carefully walked toward a group of soldiers and asked to speak to someone in command. Fortunately, I was in uniform too and was driving a state vehicle.

A sergeant appeared, and I introduced myself and explained that I was conducting a study on relocated gopher tortoises. I understood that some sort of training exercise was in progress, but I emphasized that I had to check the traps that I had set the day before. The sergeant looked me over, realized that I was determined to check those traps, and got a little gleam in his eye when I asked if it was safe for me to go out there. "Well, you can go out there ... but you might get molested." I came right back with: "Molested I can handle; I just don't want to die." Of course, I realized he was being a smartass, but I wasn't going to let him rattle me. He laughed heartily and assured me that no one would shoot me.

So out I marched into the seeming fray. In some cases, I had to work around groups of soldiers who were sitting and listening to an instructor. At one point, I was busy working up a tortoise and then a gopher frog (yes, we caught these amphibians in our bucket traps from time to time). The method for permanently marking frogs was to clip a toe (I know: kind of witch-like). I was momentarily startled when I glanced over my shoulder to find several rifle-toting soldiers staring wordlessly at me. Finally, one spoke: "You're *not* cutting off his toes, are you?" At the time, I was surprised. For Pete's sake, these were guys who are trained to shoot other guys. I calmly explained that this was the way we mark frogs; they just shook their heads and walked away.

In retrospect, I can see their point. Biological techniques are sometimes controversial, and more recent literature on marking frogs has outlined both the pros and cons of toe clipping. Needless to say, Dick got a kick out of my encounter with the sergeant. He had gotten tied up at the university or he would have been out there with me. Fortunately, that was the only day that I had to deal with an ongoing army training exercise. During later tortoise disease surveys on Camp Blanding, I had to undergo training myself, to learn how to identify and avoid unexploded ordnance. Working on military bases certainly had its share of adventure.

So what did our follow-up study show as far as tortoise relocation success? Well, we recaptured about a third of the tortoises that had been originally marked and relocated. Some were released unmarked, so we have no idea what happened to them. Many factors affected this finding. Tortoises could emigrate and move beyond our trapping area, and in fact, some do try to go home. Also, there was evidence that tortoise hunters may have been on the site at some point. And, of course, tortoise mortality must be considered as well. Part of this open, scruffy, human-altered site was later designated a UF lightning research area because (as noted earlier) Florida certainly has more than its share of lightning storms.

Permitting Woes

During the late 1980s, I unfortunately got pulled way too far into the morass of relocation permitting, which allowed me little time to conduct research. I coordinated with the lead permitting biologist in Tallahassee, Don, and together we reviewed applications for relocating tortoises. Our good cop/bad cop stance (I was the good cop) certainly got the job done, and although Don and I had an excellent working relationship, I just do not have a permitting mentality.

Those years were rough and stressful, and I was besieged daily by numerous phone calls from developers, environmental consultants (who assist the developers), county permitting folks, and others who had a stake in relocating gopher tortoises. Much later, we had full-time tortoise-permitting biologists stationed around the state, and we also convened a group of stakeholders that ranged from development lawyers to animal humane organizations.

But in the early days, it was Don and I, and when Don retired, my Tallahassee colleague Angela took the permitting reins. I also worked with agency biologists Rick, Mike, and others who were overseeing mitigation banking, where land is set aside permanently to offset the loss of habitat to development. Despite an overwhelming paperwork burden and the never-ending phone calls, I did manage to escape from my office to look at specific development or relocation sites. Most such site visits were relatively routine, but a few stand out as bordering on the bizarre. And as time went on, I was called in when a development- or relocation-related situation had deteriorated to a standoff, or where the media (another challenge) had intentionally or inadvertently blown the lid off some already contentious scenario.

In the case of the South Florida golf course mentioned earlier, the wildlife officer involved sarcastically said something like, "Just what we need, another golf course, and the hell with the gophers." I think some of that sentiment was attributed to me by the local newspaper, and I was mortified. "I was misquoted," I exclaimed to my bossman, Tommy.

"Welcome to the real world," he replied.

Fortunately, most of those crazy media moments passed by as other crises took precedence. But one small-town northern Florida paper deemed the gopher tortoise an "economic predator," and the appellation unfortunately stuck.

Two of the more bizarre permitting scenarios brought home the point that wildlife biologists also need to have expertise in human psychology. Moreover, if truth is indeed stranger than fiction, then some of these scenarios could serve as future fodder for another Carl Hiaasen darkly humorous, eco-freak novel. (Carl H. is a longtime *Miami Herald* columnist, novelist, and ardent environmentalist.) In one case, a landowner, Mr. B, whose property adjoined a tortoise relocation site, was vociferously complaining that his land was "being infested" by the tortoises and that they were killing his pine trees. (Say what?) He was calling our Ocala and Tallahassee offices frequently, and one of our regional biologists, Julie, asked me to accompany her to meet with him. The consultants who had relocated the tortoises also were going to join us for this meeting.

However, when we got to the gate, he met us there and said he would shoot the consultants if they came on his property. Julie looked at me as if to say, "I'm out of here; this guy is nuts." I felt there was more to this story, so I quickly took charge, told the consultants to leave, and asked Mr. B if we could come in and hear about his concerns. We were in uniform but still could not enter his land without permission because we were not law enforcement. Somehow, though, this situation needed to be resolved.

At first, Mr. B, an older man, was defensive and confrontational. He showed us the pine trees in question (no problem with them), and we did our best to assure him that the gopher tortoises wouldn't hurt his trees. At the very most, they would dig holes in his ground. Moreover, only a few tortoises had encroached into the edge of his land. I suggested we sit down at his kitchen table to really talk this out. Julie and I quickly learned that Mr. B's wife was seriously ill, and he was trying to carry on alone while she was hospitalized. He was stressed and lashing out at those things he thought he could control. I felt bad for him, but we had to stick to solving the "gopher crisis."

We explained that gopher tortoises, like all animals, move around; furthermore, the tortoises that came onto the edge of his land may or may not be ones that were relocated. If he could just tolerate the wild animals that shared his environs, especially those that did no real harm, he and the critters could live harmoniously. He seemed to assimilate that sentiment, and Julie and I left while we were on a high note. I can't say that we never heard from him again, because years later, our agency did get some sort of crazy letter from him. But at least we had defused the situation for the moment.

In a totally different but equally strange case, the consultant and the developer were at odds on a megadevelopment on Florida's southwestern coast. A higher-up in Tallahassee dispatched me down there to solve this apparent impasse. Time is money for developers, and time was running out before the land clearing was to commence. Poor communication exacerbated an already tense situation, and a few burrows had been accidentally bulldozed. The remaining tortoises needed to be moved immediately.

My first task when I arrived on-site was to view the situation with both parties, and I somehow ended up sitting between the consultant and the developer in the consultant's SUV. Every time the consultant said something that irked the developer, which was frequently, the developer would pinch me. As with Mr. B's case, I thought to myself, "I don't get paid enough for this craziness."

We came up with a game plan to excavate the tortoise burrows immediately, and there were two blessings associated with this otherwise chaotic and contentious scenario. One of the consultant's employees was laid back and field savvy, so I suggested he be the one to assist; and the backhoe operator, Johnny, was truly gifted in the way he could take just a few inches of soil from the burrow shaft at a time. We proceeded to excavate burrows and remove tortoises, using a long hose or poles to determine the burrow's direction and maintain the integrity of the burrow lumen or shaft. In backhoe excavations of tortoise burrows, the machine does the heavy work, and then the biologists go down into the gaping hole and excavate with a shovel. That way, there is less chance of crushing a tortoise with the backhoe bucket. And the bucket must not have sharp teeth that could puncture a tortoise. To assure the safety of humans and tortoises, communication between the backhoe operator and the biologists who are down in the eventual pit is paramount.

Of course, this atypical case couldn't go smoothly. About halfway through the day, the backhoe had mechanical problems, and we had to wait for another one to be brought onto the site. Due to some muddled contract issues, Johnny was not allowed to operate the new backhoe. The replacement operator was a disaster, and as the backhoe bucket swung just above my head, I had to put safety first. The developer was not pleased when I mandated that unless Johnny was back in the driver's seat, there would be no more excavations that day. Lots of scrambling and phone calls ensued, and, amazingly, Johnny climbed aboard another company's backhoe and we finished the burrow excavations. And I think that's when I realized that I had to get back into my research saddle one way or another and let others deal with these absurd human-fomented fiascos. Fortunately, my buddy Paul was now my supervisor, and he and I petitioned Tallahassee relentlessly to release me from such intensive permitting review and allow me to do what I did best: answer questions about this enigmatic burrowing turtle.

Team Tortoise

Although I burrowed back into my research and conducted the follow-up tortoise demographic studies as well as extensive disease investigations, the gopher tortoise relocation permitting and mitigation machine churned onward. Despite all the blood, sweat, and probably a few tears of frustration by agency staff, we just were not gleaning the desired conservation benefits. So we hatched the first internal gopher tortoise issue team, dubbed GT1, in 2004, and we were on GT3 when I retired in 2014. My comrades and I also created a comprehensive management plan for this high-profile species during that period.

You talk about a lot of work. Sometimes our team leader, Greg, and I would be on the phone honing some part of the plan for countless hours. I was therefore beyond thrilled when, in late 2007, Greg hired a gopher tortoise management plan coordinator who could oversee implementation of this vast blueprint for conserving gopher tortoises and their upland habitat. Deb quickly became an amazingly valuable asset and someone to whom I could pass the gopher tortoise baton. Sometimes I hear war stories of competition among colleagues, especially if one party feels threatened by a newcomer. Such shenanigans are not my MO; I am decidedly a team player, and Deb and I forged a strong and synergistic relationship. I truly owe a lot to her; we remain close friends even though I now live in New Mexico. She generously nominated me for my agency's highest award, and rare tears came to my eyes when the head of the agency called me to say that I had won. Truly, winning such a prestigious award wasn't even on my radar, and to win it after three decades on the front lines and as I was exiting in retirement made it all the more special.

"No Tortoise Left Behind"

Over my three-decade tenure as a tortoise biologist in Florida, I worked with many individuals who were sincerely dedicated to conserving gopher tortoises, their burrows, and their high, dry habitats. But no one was more totally immersed in saving gopher tortoises than my dear friend Ray. He was undoubtedly the most recognizable spokesperson for gopher tortoise conservation. Ray and his amazing wife, Pat, dedicated their lives to educating folks about this unique species; moreover, they bought land and set up a tortoise preserve and institute. Sadly, we lost Ray in 2010 to pancreatic cancer. When I gave a tribute presentation about his life and career, I noted that a mighty oak had fallen. His "No Tortoise Left Behind" campaign generated more than cool T-shirts; it spawned activism and gave the tortoise a voice, so to speak.

Ray was determined that tortoises not be left to struggle or die on development sites. As I indicated earlier, he assisted me from time to time in my research but also conducted his own studies. Ray wore many hats beyond educator, activist, and researcher. He was a zoologist, museum curator, ecotourism director, innovator, consultant, author, and mentor. Even knowing the extent of Ray's impressive résumé, I remained truly amazed at how far-reaching and diversified his talents and accomplishments were. He was an incredibly knowledgeable naturalist and visionary who promoted a big-picture approach to conserving and managing ecosystems. I used to tell him that I didn't know how he managed to juggle so many undertakings, often simultaneously. Of course, this juggling sometimes led to conflicts of interest or conflicts with other biologists. He was definitely an alpha male as well as a perpetual silverback (a reference to dominant male gorillas), even before his hair turned gray. His force-of-nature side caused him to really irritate folks, and I spent a lot of time trying to explain or defend Ray to my agency colleagues and others.

Ray and I were very good friends, but we certainly butted heads at times. The most notorious skirmish occurred at one of the many gopher tortoise stakeholder meetings, which were held during and after the management plan undertaking. In most cases, agency staff like myself could speak only if asked a direct question or if the facilitator requested our input. Our facilitator that day, Perran (a renowned crocodilian biologist), could be quite curt if he felt the circumstances warranted it. But as Ray droned on against allowing gopher tortoise pulling (procuring tortoises with a long PVC rod, once used by gopher hunters) to be an accepted method to relocate tortoises, I squirmed in my chair and looked at my team members who seemed to be egging me on. To their way of thinking, these were the Gopher Wars and I was Joan Skywalker.

For some unknown reason, Perran let me speak, probably because he knew that I had much more experience than Ray with the fine art of pulling. As with any tortoise capture method, pulling should be undertaken only by responsible and experienced individuals. Although few persons have this talent, pulling is still one of several methods used to relocate tortoises. Because a puller can't necessarily capture all tortoises on a proposed development site, bucket-trapping or backhoe excavation may be used to finish the task. Ray's expertise was with backhoe excavation, so he naturally advocated this mechanized capture technique. I later teased that Perran had to figuratively muzzle Ray and put me on a leash to terminate our intense exchange and counterexchange. Ray was furious, and my heart was thumping way too fast.

I'm not one to hold a grudge, so Ray and I met at an eatery shortly after this meeting, and I recall that it had a two-part swinging door like an old saloon. For a moment, I was transported back to the Old West, and there was big and backlit Ray coming through those swinging doors. Fortunately, we were able to mend fences and remain friends, despite our occasional differences of opinion on how to achieve the best results for tortoises being relocated.

One of my fondest memories of Ray was the rewarding tortoise workshop we attended together at the Mapini Biosphere Reserve down in the far reaches of Mexico in 1994. Located in the Chihuahuan Desert in the state of Durango, due south of Big Bend, Texas, this biologically rich reserve protects highly adapted but vulnerable ecosystems and their indigenous fauna. Naturally, one of the more charming denizens is the rare Bolson tortoise, a relative of my gopher tortoise. Our introduction to the Bolson tortoise field station in Mexico's remote Zona de Silencio, so named because radio waves are apparently gobbled up, was a surreal, fishtailing, nonstop (literally or you would be stuck) SUV ride through a pudding-like, rain-slick desert.

In 2005, Ray and I traveled with other tortoise fanciers to a biologist's dream locale, the Galápagos Islands. Ray had an extensive travel background and many international contacts; he was therefore able to specifically tailor this ecotour for turtle folks. It was a whirlwind ten-day cruising sojourn in the islands, in which we experienced a cornucopia of awe-inspiring wildlife. Female sea lions nibbled my fins as I snorkeled, and a subadult sea lion climbed into our dingy with us. Actually, this mischievous youngster goosed me when I was trying to climb aboard. We got to see huge wild Galápagos tortoises lumbering into a pond, nearly tripped over marine iguanas, and saw blue-footed boobies and albatrosses up close and personal.

Ray's connections got us a VIP, behind-the-scenes tour at the Darwin Station. I felt fortunate to see the infamous Lonesome George, the last of the Pinta Island tortoises, who has now gone on to tortoise heaven. And we saw the captive propagation process from eggs being incubated and reportedly kept warm by hair dryers, to charming juveniles that were destined to be released back to their home islands when they reach three to five years of age.

A few days after Ray's passing in 2010, I was training for an upcoming backpacking trip and allowing nature to ease my sadness, when I came upon a barred owl perched above the trail. Several more times as I followed the meandering path through the hardwood hammock, the owl appeared. Finally, I stopped and asked, "Ray, is that you?" Of course, I received no answer—but it would be just like ole Ray to send a message through an owl.

The Fine Art of Managing a Superb Landscaper and Landlord

So, one might ask, how do you manage gopher tortoises? Certainly, that question arose from the media as our team unveiled and then updated the management plan. The overarching goal was to restore and maintain secure, viable (reproducing) populations throughout Florida so the species no longer warrants listing. Our detailed, intensive, and thoroughly scrutinized plan addressed many, many strategies and actions under four main objectives: minimize the loss of gopher tortoises, increase and improve gopher tortoise habitat, enhance and restore gopher tortoise populations, and maintain the gopher tortoise's function as a keystone species. These objectives may sound lofty, but they are doable. And they are getting done.

One huge coup has been the restocking of gopher tortoises on Eglin Air Force Base, where I had accompanied a puller all those years ago and where I gathered tortoise blood. I endeavored to get these vast sandhills restocked after the air force restored the habitat through prescribed burning (fire is indeed the best way to manage most tortoise habitats). Alas, during my tenure, the biopolitical and military moons were just not aligned properly to allow me to achieve this important undertaking. But I am proud to say that the colleagues who took the baton from me have been able to succeed: Gopher tortoises are going back to their species' former haunts on Eglin.

There was never a shortage of tasks to accomplish on behalf of gopher tortoise conservation. We definitely had to come up with a better way to deal with waif tortoises, those of unknown provenance, who were becoming an incredibly time-consuming problem for me and my colleagues. The proximity of gopher tortoises to humans in Florida leads to issues that go beyond the obvious loss of habitat to development.

In the Sunshine State, the two most common injuries to tortoises stem from either dogs or vehicles. Being dog-bit or car-hit is not at all uncommon, and wildlife rehabilitators must then put these cracked-up gophers back together again. In fact, tortoises end up in rehab more often than some Hollywood celebs. And a number of these individuals can't go home because locality data are lacking or because they were found roaming downtown Orlando. To add to this problem, roadside gophers are easily picked up and transported to some location where they are unceremoniously dropped off. For vacationers from up North, that somewhere may be Ohio or New York. On more than one occasion, I went to the Gainesville airport to pick up a gopher tortoise being shipped back via Delta Dash. And of course, most folks can't recall the exact location where they found this roaming turtle. Education is the key here, and part of our plan dealt with helping John Q. Citizen understand that moving gophers willy-nilly does no good and only creates further problems for wildlife biologists and officers. This species is listed as threatened, and it is illegal to possess one without a permit. Although most wildlife officers might overlook a good Samaritan helping a gopher across a road, placing that tortoise in your vehicle and driving off is a horse (or is that a turtle?) of a different color.

Perhaps the most commonly asked questions by the media were, "What good are gopher tortoises, and why do we need them?" I think most wildlife biologists find such questions irksome because we understand and appreciate the ecology of plant and animal species and how they interact and fit into the natural world. It's tempting to become misanthropic and inquire, "What good are humans?" Certainly, our species has had the most impact on our planet.

But it seemed that many reporters were not overly interested in anthropogenic impacts to Florida's wildlife and their habitats. So I tried my best to explain that the gopher tortoise is a keystone species, one that provides refuge to more than 360 other species. This land turtle is an excavator, landscaper, and landlord. Quiet, charming, and innocuous, it is a model of longevity and persistence, and all it needs from us is availability of food, space, and a little tolerance. In recent years, the term *biodiversity* has been in vogue, and gopher tortoises certainly enhance biodiversity. They are the primary grazer of the dry, sandy lands, and they thereby distribute seeds. Moreover, their excavations into the earth bring nutrients to the surface.

But it's their role as landlord that is most frequently heralded. I have already noted their importance to indigo snakes, but many other animals use these burrows. Yes, there are sometimes rattlesnakes and black widow spiders, so it's wise to look around before sticking hands or head into the black recesses. Reporters and others used to get a kick out of my stories about surprise encounters with wild and domestic critters in gopher holes. Out in the Panhandle, I once encountered a skunk looking back at me from a burrow entrance, and in north Miami, I had a feral German shepherd explode from a tortoise burrow that it had enlarged for its own use.

It might be said that I was laser-focused on this ecologically important burrowing reptile species during my career; but there were other critters out there in the Florida wilds that needed attention. Of course, my colleagues were responsible for answering questions about a host of mammals, birds, reptiles, amphibians, and fish. But I would also be pulled into some fascinating experiences with species other than tortoises.

10
Rattlers and Softshells and Bears, Oh My!

The Stinky Skin Shop

Although the bulk of my research during my thirty-plus years in Florida involved gopher tortoises, I also conducted studies on the harvest of rattlesnakes and softshell turtles. Little was known at the time about the exploitation of these reptiles; I guess my supervisors knew that I was the gal for the job. After all, I had a lot of experience under my belt with snake hunters and gopher pullers. One of my more unusual stints associated with the snake study took place at the Skin Shop in the small town of Waldo, northeast of Gainesville. Contrary to whatever images that name might conjure up, this wasn't a strip joint. It was a snakeskin-processing facility, and my job was to collect data from dead rattlesnakes, much like biologists do at deer check-in stations during hunting season. Trust me when I say that this was not a task for the squeamish. And I'm not referring to most folks' ophidiophobia; instead, both the visual and olfactory aspects were decidedly unappealing, to say the least.

As I was to learn from rattlesnake dealers and collectors, most snakes brought to this facility were opportunistically killed on roads or in yards. Unlike in Georgia, there was minimal directed rattlesnake hunting. This opportunistic approach meant that snakes sometimes sat in the back of pickups for days before being taken to Waldo and thrown into the Skin Shop's super freezer. The result was that on the scheduled processing days, the thawed individual snakes gave off an odor that ranged from distasteful to downright gagging.

Now I understood why my ole snake hunter buddy Amos had called such snakes "half-mortified." I knew it was a bad processing day when one or more of the skinners (locals who had been doing this for a while) became nauseated. I always tried to position myself near a window, but the disgusting odor still permeated the long room. Before the snakes were skinned, I would measure them, count the rattles to obtain information on number of times the snake had shed, and collect female reproductive tracts for later analysis back at my laboratory. Even my colleagues were less than thrilled when I worked up the ripe repro tracts in the main lab, and these guys and gals were used to wildlife odors. It just so happened that I conducted this part of the rattlesnake harvest study shortly after Dave and I were married. Although he was very tolerant and supportive of my wildlife work, even he had his limits. I had to strip down to my undies out in the garage before entering the house and proceed directly to the shower. My uniform would have to be washed *that* evening, not merely thrown into the dirty laundry bin.

Over a yearlong period (eight data collection dates), I examined 714 harvested rattlesnakes. The majority were eastern diamondbacks, but some were canebrakes (a less common species). In most cases, these snakes had just been in the wrong place at the wrong time. During my phone interviews with collectors, I did hear stories of pets and livestock being bitten by rattlesnakes. The majority of folks that I contacted indicated that they would kill rattlers that they encountered, especially near human habitations, regardless of any potential monetary gain from selling the snakes.

As one might imagine, the Skin Shop created all sorts of rattlesnake skin curios, as well as hatbands, purses, belts, and boots. Although I personally would rather see the skins on the live rattlesnakes in the wild, I had to admit that the boots in particular were impressive.

Snake Busting

Not surprisingly, the Wildlife Research Laboratory in Gainesville received calls about rattlesnakes in people's yards. For quite a while, local snake fanciers handled these calls through an organization known as Snake Busters. In other cases, Paul, our staff herpetologist, would sometimes assist with such calls and concerns if he was available. But on one memorable occasion, I was asked to handle a tense and thorny situation at a west Gainesville residence. A large diamondback had been sighted, and the woman who owned the property called the cops. The police arrived and proceeded to try to shoot the snake, which wisely took refuge in a huge woodpile. The police then called "Game and Fish," my lab.

It was a brutally hot July afternoon. I packed up both a snake-carrying box and Pillstrom tongs (used for handling venomous snakes). I would have preferred just using a snake hook, but my colleagues urged me to take the tongs, which was good advice under the circumstances. When I arrived, the landowner and her gal friends expressed some slight disappointment that a man wasn't going to undertake this task. That sexist attitude would change shortly; besides, the cops were men and all they did was irritate the snake.

I quickly assessed the situation: The snake was still in the woodpile, and its rattling was ominous and loud. Of course, a neighborhood crowd had gathered by this time. The first thing I did was to tell the landowner that all people and dogs needed to step way, way back. I think the landowner realized that things could get dangerous, and she was able to convince at least some of the folks to move along. My heart was beating fast as I carefully removed pieces of wood using the tongs.

As I got closer, the snake's rattling intensified. It crossed my mind that a blistering hot day like this, when my heart was going ninety miles an hour, would be a very bad time to sustain a bite from a venomous snake. Painstakingly, I disassembled the woodpile log by log, and the snake was about to make a run for it (so to speak) when I grabbed it with the tongs and quickly dropped it into the snake box. The landowner and her remaining cohorts erupted into applause and hoots and now noted that it was a good thing that my agency dispatched a woman. The snake and I left, and Paul later released the impressive diamondback far away from civilization. That snake was quite lucky that it didn't end up as a belt at the Skin Shop.

Doing Hard Time at Lake Okeechobee

My next nontortoise study took me far from my high and dry sandhills to Florida's lakes, large and small. Although I did interview turtle fishermen who collected on the Central Florida lakes tucked among burgeoning development, the majority of my interviews required me to circumnavigate South Florida's huge inland lake. Towns like Clewiston are decidedly sugar country, with cane fields stretching to the horizon south of the lake. I recall staying in a rustic motel there where the TV stand seemed perpetually leaning to one side. (I guess I should be glad I had a TV.) And then there was Belle Glade, which at that time was the AIDS capital of Florida and a rough town, for sure.

Fortunately, many of my interviewees resided in Okeechobee, on the northern side of the lake. I would go to the fish camps and ask about turtle fishermen, and one name that kept cropping up was Frankie. Each time his name was mentioned, I would also hear that I was wasting my time because "Frankie don't talk to nobody." Well, I reckon there was a method to my boss's madness in sending me down to the big lake. I went to Frankie's house, and possibly because I was a woman (albeit in a state uniform), I not only had no trouble getting Frankie to talk about collecting softshells, but he showed me his catch, introduced me to his family, and basically told me more than I ever wanted to know. Frankie reinforced something that I already knew: Unlike gopher tortoises, softshells are not gentle, and they are able to snake that long neck around lightning fast and viciously bite you.

Another uber-helpful fisherman was Ronnie, who caught softshells up near Sanford, east of Orlando. He too was quite a character, and I heard the frequent refrain about loss of hunting and fishing opportunities due to development, and rich people in their big houses complaining about noisy boats on the lakes. I truly felt bad for these men and women who had lived close to nature, in the forests and on the lakes, for their entire lives—and now their way of life was being usurped by urban newcomers. This connection with nature and living off the land was, and is, disappearing, along with wild habitats and wild critters. I had no answer then, except for my slogan, "Save space for wildlife."

The ethnozoological (cultural ties to these wild places and species) aspects were both fascinating and heartbreaking, especially in rapidly developing states like Florida. In some ways, the men that I interviewed near Orlando were dinosaurs, and many of their outdoor adventures and pursuits on the waters and woods were likely destined to go extinct with time. But I admired their tenacity and moxie. The Lake Okeechobee men will probably hang on a lot longer because that region of the state is less desirable for rampant development than the coasts and the swath from Daytona to Tampa.

Fortunately for me, almost all of the turtle fishermen were cooperative, and I was able to gather insights on harvest methods, seasons, numbers taken, sex ratios, and other necessary data to help make informed decisions about how best to regulate this harvest. However, in one somewhat eerie scenario, I followed up on a contact and drove down a long, narrow road that wound back into a swampy area near Lake Okeechobee. As soon as these backwoods shanty folks saw my state truck, their faces hardened, and I knew that no good could come from my getting out and trying to engage them in conversation. I waved a quick adios and backed up until I could safely turn the truck around and get out of there. A number of my friends and even colleagues worried about me during this study. I do believe that my guardian angels were working overtime to keep me safe.

I Brake for Snakes

Especially when I first started working in Florida, my indigo snake past was still fresh, and as a result, I was asked to give talks about this imperiled species. Possibly stemming from our early show-and-tell days, presentations are generally enhanced (for both the audience and the speaker) if there is something to show. So when one gives a talk on snakes or tortoises, it's helpful to have the critter in question whenever possible.

Obviously, this guideline has limitations when speaking about megafauna and/or dangerous animals. Luckily, I had Zeus, my indigo snake that I acquired when at Auburn University conducting my master's research. Zeus was the offspring of two snakes that had been in captivity before the Endangered Species Act, so he was legal (all the more so now that I worked for a state wildlife agency). And although, as previously noted, he did once mistake my thumb for a thawed mouse, he was overall a typically gentle indigo snake that could be easily handled by the public. He was an excellent ambassador for his species and for snakes in general. Thus, he accompanied me when I gave talks on indigos in various parts of Florida.

On one of our joint road trips to Tallahassee, Zeus was in the far back of my car, quietly lounging (or so I thought) in his spacious cage. I had taken the back roads from Gainesville and was blissfully enjoying the fuchsia and white roadside wildflowers. I always carried an umbrella under my seat, because rain is often just around the weather corner in Florida. As I slowed down for some obstacle, I felt what I thought was the umbrella under my foot. I kicked it back into place and continued along at around sixty miles per hour. When I felt the umbrella roll out again, I actually looked down this time and was horrified to find Zeus casually making his way next to my accelerator and brake. How did he get loose? He had never been a Houdini before, but perhaps the motion of the vehicle prompted his restlessness. I struggled to keep him from blocking my braking motion as I shakily directed the car over to the grassy roadside.

Now I was faced with a dilemma: Obviously, he could escape from his cage—and we still had a way to go on our day trip. I gently reprimanded him (to no avail: snakes don't hear the same way we do), put him back in the cage, hoped for the best, and made it to the relatively small city of Perry, about fifty miles southeast of Tallahassee. I dashed into a hardware store and found large clips that I could use to secure both sides of the cage front. Zeus stayed put, we made it to the talk on time, and I had another story to tell about my adventures with animals. Moreover, this escapade gave new meaning to the bumper sticker "I brake for snakes," which generally pertains to avoidance of these sinuous reptiles on roadways.

Gotta Watch Those Gators

Early in my tenure, my superiors expressed some concern that I might morph (or rather be conscripted) into an alligator biologist. No worries: I was a gopher gal, but I was nevertheless occasionally pulled into gator studies or adventures. On a sultry summer's eve, my buddy Ab and I were on a quest for hatchling alligators for one of many ongoing studies of these ancient, fascinating reptiles. We were on one of Central Florida's large lakes, probably Apopka, but maybe Griffin; those details have faded with time.

What I do vividly recall is my epiphany of sorts about gators, especially mama gators. I knew that female alligators were protective of their nests, some more so than others. Most folks don't think of alligators in warm, cuddly terms, but the females are amazingly nurturing for reptiles. The mama will create and guard her large nest of vegetation, respond to the high-pitched chirps emitted during hatching, and gently carry the hatched wee ones to water in her huge, toothy mouth. She will also continue to keep an eye on those babies.

Even knowing all this, I was caught up in the excitement of the moment as we waded in toward a nest and grabbed the "yerking" (my word for their calls) young 'uns. We had left our boat back in deeper water—and now I was carrying the bucket of noisy babies as we headed out. What a glorious and adventurous way to spend a humid Florida evening. But as I waded after Ab, my reverie and bliss morphed into primeval fear when Ab turned calmly and said: "If the female comes, be sure and swing up into a tree."

Say what? I had already made the point that I am not blessed with stellar coordination. How was I supposed to get into one of the trees fast enough? And would I just relinquish the babies? Paranoia set in, but fortunately, the female guardian of her watery world allowed us to escape unmolested.

I had other experiences with alligators, small and large, alive and dead. Alligators are harvested for their hides and meat in Florida. Quirkily, I seemed to be the one tasked with verifying the gender of harvested gators during one posthunt assessment. I had to insert my finger into the cloaca (vent) and feel for a copulatory organ. The gator biologists were taking measurements, and on one massive gator that was more than eleven feet long, there was little doubt about its gender: male!

In another circumstance with much smaller gators, Ab and I were determining sex in juveniles out on a North Florida lake. That undertaking is a wee bit trickier in squirming youngsters, in that gentle pressure must be applied with thumb and forefinger on either side of the cloaca. If the juvie is a male, a penis should pop out of the vent. In one case, we just weren't sure and feared we might have given the poor babe a prolapse by squeezing down there so much. Fortunately for those wee gators, our evening of collecting data was cut short when the lightning flashes heralding a violent thunderstorm drove us back to dry land.

One other occasion stands out in my mind, one that could have turned out badly for me. Top herpetologist Paul and I were on Dee Dot Ranch, a vast private landholding on Florida's northeastern coast. We had surveyed some tortoise habitat and now were checking out a scenic pond. I casually strolled out on a low dock, enjoying the gently rippling water after being out in the hot sandhills. I recall seeing a gator in the water, but its presence only enhanced the beauty of the wetland in the late afternoon's softer light.

I must note that my buddy Paul is very Spockian, as in Mr. Spock of *Star Trek* fame; he is not one for uber-emotion or hyperbole. That made his terse warning all the more chilling when he quietly said: "Joan, come off the dock now; no questions." Paul knew I was always curious and would ask why in most circumstances. Something about the tone of his voice made me quickly pivot and head back to land. Apparently, that gator was huge and was moving in toward me as I walked on the dock that was barely above water. It was focused on, following, my every move and was not behaving in typically cautious or uninterested gator fashion. Paul worked with crocodilians and noted what I did not: This gator was potentially dangerous.

I later heard that the same gator aggressively approached and acted as if it were going to climb aboard a boat that my colleague was using to survey wading birds. Apparently, it had been fed treats and was no longer a wild, wary gator. I would suspect that our nuisance gator trappers were dispatched to remove this unfortunately human-habituated animal.

Bear Wrangling

Although I was never assigned to work on black bears because we had other biologists studying them, I was fortunate enough to be invited to assist my colleagues Elina and Walt on their bear cub survival study. So instead of Queen for a Day, I was a Bear Biologist Assistant for a Day. And what a day it was: Walt and I hunched over like apes and struggled to get through the thick sand pine scrub in Ocala National Forest. Our destination was a known bear den (in this case, more like an old downed tree). My heart was beating more rapidly than normal as I heard the female bear crashing around nearby. Most of the time, these mamas acquiesce and move off when the biologists move in closer, but there have been cases when the researchers had incredibly close calls.

Although I knew that most of these Florida black bear sows would just temporarily move aside, my adrenaline was nevertheless pumping. Walt unceremoniously reached into the den, grabbed a squalling cub, and handed it to me. He took another and off we went back to where Elina and the radio-telemetry team were waiting. To quote Walt, these bear cubs were "way up high on the cute-o-meter." I helped collect data, attach radio collars, and return the little bears to their den. Fortunately, in this instance, Mama Bear cooperated and let us complete our tasks without incident.

My job was generally so demanding that I rarely took vacations during the early days of my career. However, for several years during the 1980s, my friend Mark (of indigo snake days' fame) talked me into meeting him up in Berea, Kentucky, for an annual August bluegrass festival. We would camp out and listen to pickers 'n' grinners, literally day and night. Although it always took me part of a day to unwind, eventually this musically slower pace would seep into my brain and bones. Before or after the festival, we would go exploring amid the woodlands and hollers.

One summer, our destination was Cades Cove in Great Smoky Mountains National Park. Upon our arrival in the cove, we spied a female black bear going around to picnic tables and consuming whatever suited her fancy; in one case, it was a birthday cake. Mark and I realized no good could come from this and were trying to figure out how best to intervene when a National Park Service truck drove up. Out stepped a bear biologist whom I had met previously at a wildlife meeting. I introduced Mark as a fellow wildlife biologist and inquired how we might be of assistance. Bill, the park biologist, asked if we could help with crowd control because by now a typical summer crowd of tourists was beginning to gather. Another chance to use my "Stand back, I'm a scientist" MO.

As I was explaining to folks that the biologist was going to tranquilize the bear, a young girl came up to me and said she was Jim Fowler's daughter, as in *the* Jim Fowler, of *Wild Kingdom* fame. For the uninitiated, *Wild Kingdom* preceded Jeff Corwin's and other more recent animal shows, and Jim Fowler was typically the one who wrangled the dangerous wildlife while Marlin Perkins narrated. Coincidentally, I had also met Jim Fowler previously, at a Gopher Tortoise Council meeting where he was the keynote speaker. He's a Southern gentleman and was apparently best buds with one of my former Auburn professors. (It is indeed a small world, and the "six degrees of separation" theory seems to ring true.) Sure enough, I turned around and there was Jim Fowler and his film crew. They had been doing a documentary in the park and heard about the picnic-table-visiting bear. Now we indeed had a three-ring circus going.

The bear was catching on that maybe this wasn't the best place to be, but she had cubs that she was teaching about b-day cakes, so she couldn't just lumber off into the woods. She put the cubs up a tree, and Bill shot her with a tranquilizer dart. As luck would not have it, she went toward the one place he didn't want her to go: the creek. I was struggling to keep the ever-growing crowd back at this point, and as the female splashed into the creek, some idiot tried to take her picture up close and personal. Like an angry celebrity being harassed by the paparazzi, she groggily charged and swatted at him, just missing his head. Now it was time to get serious with this crowd. "Stand back *now*," I hollered. Jim, his daughter, Mark, and I were all trying to keep folks from getting injured.

But suddenly, my attention turned to the bear, who was well into the creek and whose head was slipping under the water. I motioned to Bill, and he motioned back for me to assist him with the bear. Into the water we went, and I grabbed her head and ears so that she wouldn't drown. By now, she was definitely quite sleepy, so several large men, Bill, and I were able to carry her ashore. This was, as they say, a "teachable moment," and I stayed with this cake-loving bear as Jim and Bill lectured to the crowd about bear biology and why it's not a good thing for bears to raid picnic areas.

After the group education, we loaded the perp bear into the bed of Bill's truck, and he stationed Mark and me in the back to keep an eye on this big bruin. As we drove down the park road, Mama Bear stirred, and for a nerve-racking moment, it looked like she might shake off the effects of the tranquilizer. "You're on your own, Diemer, if she wakes up," Mark quipped. All I could do is pray she wouldn't awake until we could get her into the large cylindrical holding cage reserved for bears.

Even after we got her secured and were able to breathe a sigh of relief, we still had another problem to address: her cubs. Thankfully, most of the crowd had moved on to find dinner because this was going to be a rather long waiting game. It became increasingly obvious that the cubs wouldn't descend on their own, so Bill carefully loaded a dart and shot it into the first one's haunch. Slowly, it released its grip on the tree and began to slide downward. We were waiting below with a fireman's net. Now the second cub was darted, and it too slid down the tree trunk. Poor punkins. I'm sure they were quite confused about this entire incident, and quite frankly, Mama Bear did them no good by trying to teach them the fine art of stealing human food from picnic tables. They would, of course, be reunited with their naughty parent, and all would be relocated far from Cades Cove. Hopefully, this capture undertaking would teach the bears that hanging with humans is decidedly not a good thing. And that is how I spent my summer vacation: Goldilocks and the three bears.

Although I was fortunate to be able to study some species intensively, and interact with others peripherally, another not-so-wild species had captured my heart from an early age: *Canis familiaris*, dogs. At one point when I was searching for a grad school, I even considered pursuing studies on coydogs, which are not coquettish pups but are crosses between wild coyotes and man's best friend. Alas, that desire didn't pan out, but there were certainly domestic canines that did become interwoven with my life.

11
Canines in Conservation and in My Heart

Lending an Ear

As I look back, I realize more than ever the significant role my dogs played in my life. I've long admired a quote by Roger Caras: "Dogs are not our whole life, but they make our lives whole." During my high school and early college years, a sweet, tan beagle cross, Randy, was my confidant and always adoring audience for scenes from *Romeo and Juliet* performed out on the family carport. I had become a thespian of sorts, especially when the 1968 movie debuted, and would deliver the opening lines, "Two households, both alike in dignity, in fair Verona where we lay our scene," to Randy when I was home from college.

When I went through the trauma of a divorce, my black cockapoo, Benji, would sit in my lap, stare lovingly at me, and fan my face with her flapping tongue. At the time, she truly seemed the only one who understood my sadness and tears. She and my black-and-white Welsh corgi, Elsa, were my hiking companions in Alabama. Benji was a comical and clumsy soul, while Elsa was agile and athletic. When we hiked along a rocky creek, Elsa would hop from rock to rock like a mountain goat. One time when the three of us were hiking along some kind of reservoir, Benji tumbled off the cement rim into the water. There was definitely a daunting drop into the water, but that didn't cross my mind as I jumped in after my little girl. As Elsa waited above us, I somehow managed to retrieve Benji and find a way back to dry ground. So I guess you could say that she saved me during a troubling transition in my life, and I returned the favor.

Intrepid Lakota and Twirlin' Merlin

If our dogs, like people we are close with, contribute values and lessons in our lives, then two of the dogs I had during my wildlife career provided character-building experiences and thought-provoking insights. From my black Lab–husky cross, Kotee (formal name: Lakota), I further learned the value of being tough and tenacious in the face of adversity, of persevering no matter what comes my way, and that growing older does allow some liberties. And Merlin, my gorgeous golden retriever, reinforced my fun-loving and comical nature and reminded me to always (like the old and wise song says) "keep on the sunny side of life," or in this case, the pure gold side.

Both of these dogs came to me through somewhat unusual circumstances. Kotee was a rescue in the truest sense of the word. In 2004, this six-month-old pup with a reverse white question mark on her chest had quite a stormy backstory. While Dave and I were at a wildlife conference in Canada, she was found wandering near my office location in one of Florida's many hurricanes that year. Taken to the animal shelter, she was granted a reprieve when our niece's colleague rescued her from being put down. Kotee needed a home, and, thanks to our niece, she became part of the Berish family.

We'll never know how Kotee came to be out in that storm, but I would like to think that she escaped rather than being thrown out into the deluge. Our vet noted that one's man's trash is another man's treasure. And Kotee has been, and still is (at fifteen), a treasure.

In her younger days, Kotee paced more often than she trotted (i.e., she moved the feet on one side of her body together rather than alternating), and she also would take off at a moment's notice if given the chance. Pacing down a road or trail gave her a rolling, butt-swinging, somewhat comical way of moving. These characteristics are not uncommon in huskies. And her haircoat isn't that of a sleek Lab; instead, it's more like the thick and unruly fur of a wild canid. I have heard that a dislike of water may also be manifested in huskies, but I surmise that Kotee's disdain for rain may be due to her facing nature's wrath in that hurricane. Kotee is the only canine kid that made our pioneering move to New Mexico with Dave and me in 2014, and amazingly, she absolutely loves snow; so maybe she is a husky after all.

Kotee has always been delightfully quirky. Back in Florida, one of her unusual habits was to scoop up a cicada in her mouth during walks and proceed to trek down the road, with the cicada making a deafening and disturbing racket inside her mouth. If we tried to pry her mouth open to get the cicada, she often just gulped it down. Fascinated by more than noisy insects, she has always loved watching animals on TV, and when the charming show *Meerkat Manor* was on Animal Planet, she would hear the opening music from the other room and come charging into the living room, rise up on her hind legs to get a better look at the elevated screen, and growl menacingly at the adorable furry munchkins that dared enter her domain.

Merlin's background was completely different from Kotee's hurricane rescue. Merlin was born with a silver (or rather make that a golden) spoon in his mouth. In 2007, we were happy just having Kotee and weren't looking to give her a "brother." But one spring evening, Dave came riding home on his golf cart after having three white Russians (the drink, not the people) at the club. As he cut through a nearby neighbor's yard, he stopped abruptly when he saw her holding a wee pup. When he admired the little fluff ball, our neighbor asked if Dave wanted this eight-week-old, purebred male. Being somewhat inebriated, he said yes and brought the pup home.

Neither Kotee nor I was ready for a new addition right at that moment, so I had him take the pup back with the caveat that we would discuss it (more like a pro-and-con list) and let our neighbor know the next day. Even though the cons outweighed the pros, primarily because I was so incredibly busy at work, we adopted that adorable little puppy.

Kotee was beyond indignant and growled menacingly each time the pup got into her space. At first, we wondered if this was going to work, having a three-year-old tough-as-nails dog and a new, naïve pup. Fortunately, Merlin was more than happy to let Kotee be the alpha and eventually won her over with his innate charm. As he grew older, they played rough, but he had enough sense to flop down on his back and let her pretend she was getting him by the throat. Sometimes, I did yell out the back door, "Don't kill your brother!" In an interesting twist, Merlin imitated his older sister and began pacing rather than trotting. His exaggerated, butt-rolling, motion was really quite comical, to say the least.

Both dogs loved to eat (a retriever characteristic), but Merlin was totally obsessed with food. It didn't matter if I had been gone five minutes or five days; he would greet me, run immediately to his food bowl, and do several pirouettes. For that behavior, we bestowed the nickname "Twirlin' Merlin." As Merlin matured from gangly pup to magnificent adult, people would stop us on our walks to comment how gorgeous he was. Tactfully, they also complimented Kotee because she's a black beauty. But Merlin was the one with the plume of a tail, and he brandished it like a foppish musketeer.

Dorothy and the Cowardly Lion Meet the Serpent

Two notable Kotee/Merlin stories stand out from the rest. In the first, Kotee definitely displayed her intrepid nature. Dave and I were at a family wedding in Michigan when our dog sitter, Amber, called to say that there had been a close encounter of the terrifying kind with the largest diamondback rattlesnake she had ever seen. She had just returned from a dog walk on a drizzly fall eve, when Kotee darted out the doggie door and barked furiously. Much to Amber's horror, Kotee was now playing mongoose with the snake, which rattled ominously and struck at her. Merlin took one look, ran into the house, and hid behind the couch. Somehow, Amber managed to pull Kotee away from the snake, which was bleeding from where her teeth had made contact. The mystery of why this large snake was in our yard was somewhat solved when Amber saw a dead squirrel. Apparently, the rattler had struck the squirrel and tracked it through the fence into our yard. Amber threw the squirrel back over the fence and hoped the snake would follow, which it apparently did.

High Jinks Gone Awry

The other Kotee/Merlin story resulted in considerable challenges and character building for me. Merlin epitomized his breed's joyful and exuberant nature, but he was also a rather clumsy canine. In December 2012, his antics resulted in a disastrous accident for me. It was the day after Christmas, and I returned home from work early so that I could walk the dogs on this pleasant winter day. Both canine kids were in high spirits. We were only fifty yards or so from my house when the three of us began to bounce and skip down the street. Merlin couldn't contain his enthusiasm: He looked at me as if to say, "I love you, Mom," and then rose up into the air, hitting me like an NFL linebacker. I came down on the asphalt, directly on my left hip. Had I landed three feet farther to the left, I would have been on grass.

I immediately knew that something was dreadfully wrong. Before I could assess the damage, I had to somehow wrangle the dogs. Merlin was ecstatic: "Mom's on the ground to play with me." Kotee realized that we had a problem: "You idiot," she seemed to say to Merlin, "Mom's hurt." I saw no one else out and about at first, but then I spied a neighbor way down the street. He was outside with his glass of wine, and when I tried to summon him, he waved back. I gestured more strongly, and he finally sauntered over. He was horrified when he saw what was really going on, and he set his wine glass down so that he could take both leashes and put the dogs back in my house. While he was doing so, I unceremoniously scooted across a yard to my next-door neighbor's door, dragging my left leg. My neighbor later said she wondered what in the world I was doing at first, but then of course she came to my assistance and called an ambulance.

The bottom line was that, at age sixty-two, I had indeed broken my hip. That resulted in surgery, several totally miserable days in the hospital at the holidays, and months of healing and physical therapy. I have three large screws in my hip permanently. I knew that my Scandinavian heritage and genetics predisposed me to osteoporosis—but I wasn't prepared for this. One of my dear colleagues reflected his sense of humor when he dubbed me "the high-risk dog walker." I of course didn't blame Merlin; he would have never meant to hurt me. But now I had to summon the Scandinavian (perhaps Viking?) tenacity and fortitude to come back from this serious injury.

I truly attribute my return to "normalcy" to my physical

therapist, Marty. He had been my PT on a number of other occasions from past work-related injuries; moreover, he had worked with the University of Florida football team and was considered one of the top PTs in Gainesville.

Marty worked me hard, but it's what I needed. My office colleagues were also supportive but tough: "You're limping," Elina and Woody would note as I made my way down the hall with my walker, causing me to straighten up and try to walk more normally. Although I was forced to attend professional meetings on a walker and then use a cane for a while, I did eventually resume my fieldwork without assistance from those devices. I couldn't squat to pee in the woods for nearly a year, and that's a royal pain when you are afield. I knew I was really back in the saddle when I could perform this bodily function out in nature.

Although Merlin of course couldn't understand the havoc he had wreaked in my life, he would stoically sit by my bed so that I could pet and adore him. And adore him I did. I would sing to him, mostly the poignant song "Little Boy Blue," recorded by country singer Lacy J. Dalton, about the omnipresent love for a son. Merle was my little boy gold. At the time, I didn't know that he would be leaving us before 2013 was over, at the way-too-early age of six.

Merlin's purebred heritage proved to be his undoing. Unbeknown to us, golden retrievers are apparently subject to a number of maladies, including hemangiosarcoma, an aggressive and deadly cancer of the blood vessels. By the time the Gainesville specialty clinic veterinarian determined the diagnosis, it was too late. We were heartbroken but resolute that we would not let him suffer or "crash and bleed out." So three days before Christmas, our dear friend and vet Lori came over to peacefully help Merlin travel over that oh-so-sad Rainbow Bridge. Dave and I stayed by his side as Lori tranquilized him and then gently administered the drug.

Kotee also came to say her good-byes to her sweet soul of a brother. I had put on some soothing Native American flute music to help us all through this difficult morning, and as soon as Merlin passed, the song "Amazing Grace" (which I didn't even know was on this CD) played, and we just looked at each other with a mix of awe and profound sadness.

A Nose for Data

Another dog that had a significant impact on my life and perhaps could have positively impacted wildlife was Stormin' Norman. In one of those aha moments, it recently dawned on me that of the three large-breed dogs that I had during my wildlife career, he might have been a great conservation canine.

The use of conservation K-9s is a burgeoning field that is yielding incredible results on behalf of wildlife and wild habitats in many parts of the world. There is also a symbiosis involved, a real win-win, where both dogs and wildlife benefit because many of the canine candidates come from shelters. In some cases, the best conservation dogs are so-called bad dogs. These are the dogs that people dump in shelters because of their extreme energy, their intense drives, and other characteristics that render them less desirable as house pets. But these are the very qualities that make them outstanding conservation dogs.

These amazing dogs are used for a host of projects: finding grizzly bear and wolverine scat (for locality and DNA data); detecting invasive species like zebra mussels and noxious weeds; locating elephant ivory and other poaching-related contraband (including guns, snares, and ammunition); and finding cryptic and/or rare species like certain frogs and toads. To quote Megan Parker, cofounder of Working Dogs for Conservation, "Dogs are the only species on the planet that can be bothered to bring us stuff and bothered to tell us what they know."

It's no secret that dogs have truly amazing olfaction. They are the Olympians, while we humans are pitifully inferior (yes, even perfumers fall short). It has been said that a dog could detect a teaspoon of sugar in a million gallons of water (approximately two Olympic-sized pools). For example, Tucker, a recently retired black Lab from the University of Washington's Center for Conservation Biology, could sniff out orca (killer whale) feces floating on Seattle's Puget Sound, even if it was a half mile or more away. And there are scent-detection canines that can even find the tiniest droppings, for example, from endangered species such as Southern California's

Pacific pocket mice and Mount Hood's caterpillars of the silverspot butterfly.

Closer to home for me, dogs are now being used to locate indigo snakes, and my buddy Ray had trained his yellow Lab, Sandy, to help find gopher tortoises. Similarly, conservation canines have been used to locate desert tortoises in the vast reaches of the Mohave Desert. Farther afield, but still tortoise related, a Belgian shepherd named Brin was used to find critically endangered geometric tortoises in South Africa.

Sniffer dogs are not the only canines being employed for conservation efforts. Certain breeds have a long history in Europe and are ideal for reducing human-wildlife conflict. In Namibia, Anatolian shepherds and Kangals (both Turkish breeds) guard livestock from cheetahs and other predators. This is one of the many successful programs conceived by Laurie Marker, founder and executive director of the Cheetah Conservation Fund. The pups are born and raised with the livestock they will be protecting and, in effect, become the so-called smart members of the herd. Here in the American West, Karelian bear dogs, originally from Finland and western Russia, are being used to haze and deter bears from coming into problematic and sometimes deadly contact with humans. These beautiful dogs, with their striking black-and-white faces striped like a skunk, basically harass bears by getting in their faces like pesky gnats.

A K-9 Force of Nature

So how did Norman come into my life, and why did I think he could have possibly become a conservation dog? Unexpectedly in 1996, this large, rangy, two-year-old black Lab cross came winging his way to us from France. Little did Dave and I know that we were entering the wonderful, challenging Stormin' Norman chapter of our lives. For those who have read John Grogan's charmingly captivating book *Marley and Me*, suffice it to say that Norman was the dark-hued doppelgänger of that rogue, yellow Lab Marley.

A half-breed (Labrador and golden retriever), he was whelped in the south of France and subsequently raised there by my brother-in-law Ken until his "terrible twos." His moniker Stormin' Norman came from the much-heralded General Schwarzkopf, and it apparently fit both his actions and personality. Our sister-in-law Janette swore that Norman was trouble from the get-go (perhaps an early sign that he might have made the grade as a conservation dog). Even as a half-grown pup, he bowled her over, escaped his confines, and roamed the French village. When my brother-in-law was transferred to England for work, he sought to accommodate his beloved and energetic Norman in one of his eight siblings' homes because he knew that this fractious dog would not fare well under England's mandatory six-month quarantine.

Along the lines of "many are called, but few are chosen," Norman came to live with me and Dave, where he reigned supreme for eight years. There was undoubtedly a synergy between Norman and Dave and a propensity for getting into mischief. At times, I joked that they were both stuck in a neotenic (larval) state, refusing to grow up.

Although I didn't realize it at the time, Norman had many characteristics of a conservation detection dog. I loved him beyond measure—but he certainly could have been classified along the continuum of "bad" dogs (mischievous but not malicious). His energy was relentless, and his prey drive was off the charts. I suppose it would figure that Norman was fascinated by turtles and tortoises, but not in a good way. "Walkin' Nylabones" is what my vet friend Lori called these shelled reptiles because they seem to be "dognip" to canines, both wild and domestic.

In one case along a woods trail, I had let Norman briefly off leash to stretch his legs; I quickly learned that having Norman off leash was not a good idea. He ran ahead of me along a creek bed and promptly dove down to flip over a cooter turtle so that it couldn't escape. Had I not been with him, he would of course have proceeded to gnaw on it. Another time, I noticed that he was acting strangely in our backyard, which bordered a nature preserve. He looked . . . well, guilty and secretive. "Something's up," I thought, so I started looking around, and there, stashed in a corner, upside down, was a gopher tortoise. Now how would it look if the state gopher tortoise biologist's dog set aside a gopher as a chew toy?

I reprimanded Norman, climbed the fence with the gopher tortoise in hand, and gently took it back to the one burrow I could find. The hardwood hammock behind my home was not good tortoise habitat. It was thick and shaded, so this tortoise was likely just passing through and scooted under the fence to almost meet its demise via my dog's jaws.

I now wish that I had further explored options for employing Norman's drive and obsession into a useful pursuit. But my personal and professional worlds were so incredibly busy that I could enjoy interacting with him only as my canine kid.

Like most Labs, Norman had a tendency to be ruled by his stomach. And as a scientist, I saw firsthand the concept of Pavlov's dogs. There happened to be an old Southern gentleman, an original owner of our golf community, who loved driving his golf cart around during the late afternoons when I was walking Norman. This kindly man, Norwood, loved dogs and carried biscuits with him on his golf cart to give to any canines he encountered. Norwood was especially enamored with the retrieving breeds, and his colorful comment was always the same: "Them Labs won't bite butter," alluding, I presume, to the breed's gentle nature. He developed a particular fondness for Norman, and of course Norman reciprocated.

However, this fondness got to be a problem for me because if Norman saw Norwood, I was unceremoniously pulled toward the incoming golf cart. Norman sometimes nearly pulled my arm off when he saw a squirrel or cat or, in this case, Norwood. Images of a bug-eyed cartoon character being dragged or yanked aptly convey some of my walks with Norman. Norwood never seemed to comprehend that I could be hurt by my powerful canine. Their interactions waned a bit after one encounter when my brother-in-law's yellow Lab, Chelsea, was also present. Norman was obviously jealous about sharing biscuits with his cousin and chomped down on Norwood's fingers. Fortunately, no skin was broken, but Norwood learned that, under certain circumstances, those Labs will indeed bite butter and anything else in the general vicinity of their mouths.

Although the cooter in the creek bed incident taught me to keep Norman on leash (often at my own potential peril), Dave and his older brother Bob saw little reason to curb Norman's enthusiasm.

In one memorable incident, Bob felt that the canine exercise task would be accomplished so much more efficiently if Norman just ran around loose in an undeveloped lot. Norman, not surprisingly, had other ideas and bolted through a small gap in a barbed-and-hog-wire fence in pursuit of a deer. White-tailed deer and wild turkey sightings were common in the adjacent nature preserve, but Norman rarely got the opportunity to interact with these forest denizens, having to content himself with looking and not touching. Ever the obsessed predator, Norman loved to chase critters. And now Uncle Bob had given him this wonderful and rare opportunity.

Upon being informed of my missing canine kid, I chewed out Bob and went in search of my wayward Lab. This wasn't his first escape. Back in our earlier development, he had cleared a neighbor's six-foot privacy fence (in and out) and had even taken a hit from an electric wire on top of a chain link fence during a thunderstorm (which he abhorred). In fact, he often seemed to be channeling Houdini, but Dave always claimed that he would return. Much to my relief that hot day, he did eventually gallop noisily back up the road, like a big-boned, black gelding.

Thinking that, once again, Norman had survived his escape (or, in this case, intentional release), I retreated to the shower to wash off the sweat accumulated in my pursuit. From beneath the cascading water, I heard a decidedly male scream (not typical in my house) and emerged dripping from the shower to find an extremely agitated Dave and a nonchalant Norman (still reveling, I suppose, in the joy of his unexpected romp). Apparently, Dave had gone to pet Norman's shoulder, and his hand went into a clean, but amazingly deep, sliced hole in our dog. At that point, chaos ensued, as some family members tried to pile into the Jeep with the dog, and others tried to figure out how this injury occurred.

The mystery was solved when Norman's hair was found on a thankfully nonrusted strand of barbed wire. Naturally, it was a weekend, meaning that the fast-driven trip across Gainesville was to a high-priced emergency veterinary clinic. And when Uncle Bob saw the bill for suturing Norman's shoulder, he immediately deemed this a bona fide "stupid pet trick." In fact, that notation was later made on the check he wrote to reimburse me for his momentary lack of judgment.

Another one of Norman's misadventures happened when Dave unwisely let him race down a suburban path. It just so happened that a bicyclist was on that same path, and Norman gave chase. The rider pedaled faster and yelled and kicked (just what a predator loves: prey that entices), and although Dave hollered that Norman was not vicious, the rider of course wasn't buying that sentiment. Norman had exceptional eye-to-mouth coordination and thought this chase was a really fun game, so he leaped up and gently (?) pinched the biker's leg. "Your dog bit me!" the man yelled at Dave. Dave ran up and asked to see the bite. There was no mark on the man's calf. The episode would have likely ended there with most folks, but this particular bicyclist was a lawyer and threatened to sue. "You can go ahead and try to sue me, buddy," Dave replied. "But I don't see anything." Luckily for us, the lawyer let the matter drop.

Norman's reign ended with a strange affliction, megaesophagus, where food goes both ways in the upper digestive tract. We were able to keep him going for a while by creatively feeding him in vertical position, having him stand on his hind legs with his front legs on a chair. Sadly, pneumonia is often what takes down dogs with megaesophagus because they aspirate food or water.

One hot July day (actually, it was the French Bastille Day), Norman wouldn't come in and seemed to want to retreat into the forest. I think he knew it was time, even if we didn't. We rushed him to our country vet clinic, where he passed away despite our vet's efforts to save him. Even our friends and family shared our grief over the loss of this amazing dog whose life and legend spanned two continents.

So what lesson did I learn from this larger-than-life canine? Norman taught me to be even more free-spirited and adventurous than I already am; yes, there are rules, but they sometimes need to be broken. He and all my other dogs have magnified my life in ways that really are beyond description. And that is one of the many reasons why these tail-wagging critters will always have a place in my heart.

3

Modifications to My Home Range, Both Temporary and Permanent

Life is either a daring adventure or nothing at all.

—HELEN KELLER

12
Misadventures on Desert Trails and High Seas

Yin and Yang of Adventure Travel

Much has been written over the centuries about the benefits and joys of traveling and adventuring, of shedding one's old skin like an indigo snake or a birch tree, and growing a new one. I have always been a proponent of the expansion of mind, body, and spirit that results from vacating one's home burrow and wandering far afield. And like most wildlife biologists who spend a majority of their time outdoors, I also sought nature-related adventures even during my so-called time off. But any traveler, whether visiting the grand cities of Europe or heading far into the backcountry, has to expect a dollop of discomfort or inconvenience, sometimes seasoned with moments of high anxiety. Flight fiascos, lost luggage, hotel hassles, terrifying traffic, having one's pocket picked, and other unsavory occurrences can plague more traditional vacations, but outdoor sojourns add a different layer of potential misadventure.

In my case, for example, a twisted ankle while stumbling around in the dark to pee short-circuited a Superstition Mountains backpacking trip in Arizona, requiring my friend Mark and me to find an alternative route out and to hitchhike back to civilization. Unforeseen morning delays and extreme 106-degree Texas heat turned a high-noon ride at a dude ranch into a dizzying jaunt through hell, so much so that I slid down from the saddle at the end, literally staggered into the cantina, and collapsed on a thankfully well-placed couch. And in Arizona's canyon country, my horse slipped and slid on the slickrock trail so frequently that I can honestly say that "I've been through the desert on a horse with no ... coordination."

Indeed, at times, Mother Nature can be a trickster, a bruja

(witch), or worse. Yet many of us crave outdoor adventures, not only because they offer thrills and excitement but because they can be as relaxing as watching the sun rise while sipping tea on a cool Sonoran Desert morn, or as awe-inspiring as hiking through a dazzling display of violet-blue lupines and crimson paintbrush at the foot of Mount Rainier. Yes, I've been both blessed and seemingly cursed while riding, hiking, backpacking, camping, and sailing. But these adventures—and misadventures—have enhanced my life and hopefully strengthened my character.

Camping in El Pinacate

Those of us with extreme wanderlust and an insatiable curiosity about the natural world are easily tempted to venture into new and exotic locales. So in 1991, when my adventurous buddy Mark introduced the idea of traveling to an enticing and remote part of Mexico, I was intrigued. Mark had been researching this area, especially after learning that a joint contingency from the Nature Conservancy in Tucson and El Centro Ecológico (zoo/botanical garden) in Mexico would be exploring the lava fields, craters, and rolling dunes of El Pinacate y Gran Desierto de Altar (now a biosphere reserve). The various biologists and naturalists would be joined by an author, Peter Steinhart, and a photographer, Tupper Ansel Blake, who were working on the book *Two Eagles/Dos Aguilas: A Natural History of the United States–Mexico Borderlands* (published in 1994).

Located in the Sonoran Desert of northwestern Mexico, immediately south of Arizona and east of the Sea of Cortez (the Gulf of California), this volcanic landscape is noteworthy for its dramatic combination of desert landforms, its habitat and species diversity, and its visibility from space. The human history is also fascinating: early hunter-gatherers, indigenous Native Americans (a band of O'odham [Sand Papagos]), explorers and scientists, and NASA astronauts sent there in the late 1960s to train for lunar walks. As if extensive black and red lava flows adjacent to North America's largest active dune field weren't compelling enough, this truly awe-inspiring landscape also has ten otherworldly, large, deep, and nearly circular maar (steam-blast) craters.

This would be the first of three trips that Mark and I took to El Pinacate, always in the early spring because temperatures later in the year are too intense (dare I say dangerous?) to enjoy the beauty. This initial trip definitely had fewer misadventures. Certainly, camping as part of an international group had its advantages, both in terms of safety and enlightenment about this unique region. For example, we captured a species of horned lizard that was not known to exist in that specific location of the vast sea of dunes, making our finding a range extension. If the Mexican biologists (one of whom was a desert tortoise researcher) had not been with us, Mark and I would not have recognized this new scientific contribution.

As we hiked far out into the dunes, I was taken with both their sensuous beauty and the perseverance of the flora and fauna. Especially impressive was the *ajo* or desert lily, a lone and stunning botanical wonder with creamy, trumpet-shaped flowers. An added bonus of sharing this experience with our new amigos south of the border was their evening merriment of song and guitar music around the campfire.

On this first trip, I learned to appreciate walking on the smooth, ropy lava (*pahoehoe*) or crunchy cinder and to avoid the *aa* (sharp and rough) lava. But descending into MacDougal Crater, one of the immense geological bowls, did prove both daunting and thrilling. Moreover, Mark tends to be a "full-speed-ahead hiker" and is blessed with the balance and agility of a mountain goat, so he nimbly charged down the steep, rocky slope that was littered with sharp teddy bear (a misnomer, for sure) cacti.

I, not being endowed with either noteworthy balance or coordination, had to resort to employing a crab-like gait, leaning over backward, extending my legs, and using my arms to avoid falling. I was terrified at times on that seemingly long descent but did eventually reach the bottom. My legs were like gelatin due to using my quads so extensively. How was I going to climb back out? Of course, Mark was nowhere to be seen, and I later learned that he merrily explored the bottom, didn't wait for his companion (me), and easily climbed back to the top.

Fortunately, I wasn't totally alone in this remote yet totally fascinating crater: the photographer's wife was still down there, and she and I marveled at a stately and incredibly tall saguaro cactus that stood like a lone sentinel. I hiked all around the crater floor until I thought my legs would not fail me on the climb out (I've always been better at going up than going down) and then used my arms to pull me up from rock to rock, all the while avoiding both the ankle-biting and appendage-impaling cacti. I definitely pumped adrenaline during that climb. I'm sure I gave Mark some grief for taking off like a gazelle, but that's just his way. He would have eventually come looking for me if I hadn't returned in a timely fashion. Besides, he knew that others were descending and ascending into the crater on that gorgeous spring day.

After I had rested up from my extreme exertion, I bid adios to the folks who were heading home. While Mark roamed and photographed, I strolled north toward the imposing Hornaday Mountains. I must have blissfully wandered and meandered like Prissy in *Gone with the Wind* before I realized that the sun was descending more quickly than I had anticipated, and then I had some anxious moments as I tried to find my way back to camp amid a maze of intersecting Jeep paths. Getting lost in that remote desert would not have been a desirable adventure, to say the least.

The rest of our journey through this landscape that has been compared to the surface of Mars transpired without any major mishaps. We were thrilled to see a rare Sonoran pronghorn antelope but less thrilled when a violent spring storm dumped torrents of rain in the wee hours, flooding our campsite and eventually causing us to abandon our tents and pack up our camp. Hoping for better weather, we tried to ascend the prominent Pinacate Peak the next afternoon, only to be turned back by a cold, windblown rain and dauntingly jagged *aa* lava.

Cactus-Fused Fingers in Old Mexico

Alas, my second journey into the Pinacate, in 1995, gave the term *misadventure* new meaning. Our pals Dan and Elizabeth joined Mark and me on this desert sojourn. We camped within view of a stunning stand of ocotillo, a tall shrub with cane-like, spiny stems that were topped by brilliant crimson flowers during the desert spring. And the stars that first night were, well, splendidly stellar.

This time, we took a rough track way out into the lunar landscape along the edge of the dunes. We were looking for the remote Moon Crater, but I'm not sure we found it because the rim blends into the lava. We did a long hike to the dunes, where Elizabeth and I were lured by their seductive beauty and were chastised when we ventured far beyond where the guys had hiked. A vehicle mishap that far out could be dangerous or even deadly, and Mark wanted to get us back to camp before darkness descended. Fortunately, no misadventures occurred during the dune exploration. But it was much hotter than during our first trip, and the searing spring heat meant that we needed to retreat into the sparse shade cast by the vehicle or by the meager ironwood or palo verde trees later in the day.

But it wasn't the Pinacate per se that created problems for me: It was my own body. I was forty-four at the time and had been experiencing some irregularities in my menstrual cycle. My periods seemed to arrive when least expected. Because I had just had a period, I was surprised (and not in a good way) when I seemed to be having yet another full-blown event. I hadn't carried many tampons with me out to the desert because I wasn't expecting to need them. We were a long way from civilization and a drugstore, so once the tampons ran out, I had nothing to stop the flow.

I finally had to confess to my compadres, and, luckily, Elizabeth had brought a huge box of Kleenex. Let me just say that wads of tissue in one's underwear are hardly comfortable, but desperate times call for desperate measures. I carried on with hiking and exploring, but things became dicey one night. I had to get up, hike away from the tents so as not to awaken the others, squat down, and bleed in the desert. This must have been the way it was for early native women who inhabited this remote landscape. Even with a flashlight, I had additional moments of high anxiety when I couldn't find the tents: I had wandered a bit to get away from camp. I wasn't about to holler for help or send the flashlight beam high into the tents, so I kept the beam low and walked a zigzag pattern until, thankfully, I saw the brightly colored camping tents. The next morning during the predawn, coyotes seemed to surround our camp with their yipping, and when I confessed about my nocturnal hike, the guys teased me endlessly about baiting the coyotes.

My bodily malfunctions were to cause me more grief. On this particular windy spring day, we were exploring Molina Crater. It's shaped like a three-leaf clover and is not nearly as daunting as MacDougal, so we split up and hiked wherever we wanted along the rim or down into the three-lobed bowl. I was wearing a small backpack to carry water and, of course, Kleenex, and I left the others to attend to my needs. I had just finished up and arose from a squatting position when the wind caught the dangling strap of my backpack and blew it into one of the pervasive cholla cacti. At least one of these species is known as jumping cholla, because the joints disengage so easily from the cactus and attach to anyone or anything within reach. Crazily (even for me), the strap snapped the cholla so forcefully that the small cactus joint with its insidious spines literally fused my forefinger to my middle finger. And there I was, alone, with my fingers painfully fused; thankfully, I had at least pulled up my pants. There was no way to remove the cactus spines without assistance, so I walked around in serious pain until I saw my companions ascending from the crater and could summon them.

Even Mark and Dan, who generally teased me unmercifully, were both appalled and sympathetic. They got out the forceps (which we carried expressly for this purpose) and, as gently as possible, pulled the interlocking spines from my fingertips. Tears came to my eyes as Mark finally disengaged the last spine. Fingertips are notoriously sensitive, as anyone who has been pricked for blood testing can testify. Needless to say, I was much more careful about the location of my posterior and the rest of my body when I had to squat down in the desert the next time.

Eventually, we did make it out of the Pinacate, and I was able to procure the necessities to get me through the rest of the trip. Unbeknown to me at the time, I had developed endometriosis and would undergo a hysterectomy the next year. Truthfully, especially after the problems it posed in the Pinacate, I was quite glad to be done with this bodily function that rarely seemed to be convenient, especially for a field biologist.

Slow Day at the Border

My third, and final, trip into the amazing Pinacate occurred in 1998. And once we actually got into the lava-strewn landscape, our journey was as awe-inducing as ever. This time, it was just Mark and me, and we were entering Mexico in Mark's SUV with Virginia tags. Right before we reached the Arizona-Mexico border at Sonoyta (just south of Organ Pipe Cactus National Monument), I had dropped a postcard into a mailbox. The card was to my colleagues at the Wildlife Research Lab back in Florida and blissfully noted that we were having a marvelous time in the Arizona desert. Little did I know that our fun would quickly dissolve at the border.

It was a Wednesday afternoon and apparently a slow day at this relatively small border entry point. Whether it was because the Mexican border guards were bored, or whether Mark's non-Arizona license plate made them think that we were from a land far away (therefore no local political repercussions), the individuals on duty that day harassed us unmercifully. We were shocked. This had never happened before, and we had entered at that border several other times without incident. For those who haven't experienced this, let me just say that this kind of treatment goes way beyond ruining the day: It is quite terrifying. Once you drive beyond the US border, you are in Mexico and subject to whatever the armed guards want to throw your way. It's irrelevant that you can look back and over to the other side to see the US Border Patrol officers.

Although Mark and I calmly tried to explain that we were going into the Pinacate to hike and camp, the guards kept saying that we were just going to Puerto Penasco (Rocky Point, a spring-break and otherwise vacation destination on the Gulf of California, sometimes called "Arizona's Beach"). They kept searching our vehicle and, even after seeing camping gear, didn't believe that we would be camping. They threatened to confiscate Mark's vehicle, which was indeed worrisome. They separated us, and one dark-eyed (and I don't mean that he had brown eyes; rather, his eyes were vacuous and almost evil) guard got right in my face to intimidate me. I was scared, but I tried to hold my ground and couldn't resist asking him if he would want this kind of treatment for his wife, or daughters, or sisters.

Truly, if I had been with my husband, Dave, I might have been locked up in Mexico for a long time. Dave hasn't traveled extensively and expects things to work in third-world countries like they more or less do here in the United States. When I got home much later, he wanted to know why I hadn't told the Mexican guards that my old friend and neighbor Doug was now head of the US Border Patrol. Thankfully, Mark remained calm and compliant, and, finally, even in the presence of obvious signs indicating that bribes were illegal, he surreptitiously slipped a one hundred–dollar bill on the front seat. Suddenly, all was well, and we were allowed to go on our way.

We had been detained for quite a long time, and we were understandably traumatized. We made it to a campsite at the aptly named Red Cone (a reddish, volcanic formation) late in the day and were determined not to let that absurd and unnecessary border incident ruin our trip. So we hiked up into the saddle between Pinacate Peak and Carnegie Peak and navigated over into some sharp *aa* lava on the way down. Talk about watching where you place your feet. A thrilling sighting of a western diamondback rattlesnake further reinforced the need to watch where we stepped. Being snake-bit in this remote country could be fatal.

Even though I was quite tired from all the human-inflicted stress and physical exertion, sleep did not come easily that first night. Later in the trip, we camped beneath a full moon and next to the rim of MacDougal Crater, and I arose in the wee hours to marvel at the softness of the light on the saguaros and other cacti. It was a rare natural majesty that I was witness to, and despite the drama trying to reach the Pinacate this time, I felt privileged to be there.

We also encountered a family of Papago (Tohono O'odham) Indians, appropriately at Papago Tanks (a series of water holes), and enjoyed chatting with one of them about their deep ties to this land. The tanks that serve as rain catchers are vital to humans and beasts in this stunning but harsh environment. And the cholla cactus that had painfully pinned my fingers together produces buds in April that have been described as tasting like the brilliant green of spring. The buds are harvested, used in salads and other dishes, and also dried for year-round use.

The tall, stately saguaro cactus produces its crimson fruit just before the rainy season, and long poles made of saguaro ribs are used to reach the fruit at the apex of the cactus. Syrup, jam, and wine are rendered from this Sonoran Desert icon, and there are strong cultural ties to the harvest. Mark and I were already aware of the tribe's deity, I'itoi, or Elder Brother, whose domain includes the caves at the base of the Pinacate peaks as well as Baboquivari Peak in southern Arizona. The natural, cultural, and spiritual connections to this lava-strewn desert are fascinating and profound.

We continued our explorations and luckily avoided mishap when Mark insisted on climbing up on one of Phillips Buttes, a cinder cone with extremely steep slopes and a treacherous forest of teddy bear cholla cacti. I tried to follow him, but wisely decided to remain below, and asked him what I should tell his wife, and my amiga Cathy, if he failed to return or was injured. Fortunately, he ascended and descended safely. We also hiked up to the rim of the formidable Sykes Crater on an extremely windy day. In one mystical moment, a raven appeared over the crater, relatively close and at eye level with us, croaking continuously as if to send a message. Suddenly a dust devil (whirlwind of dust or sand) came at us along the rim trail. When it had passed, the raven had vanished, gone with the wind.

Although the Pinacate has been made a bit more civilized over time, with a visitor center and more rules and regulations (for example, no hiking into craters), it's still a wild, wondrous, and magical locale that richly deserves its designation as a biosphere reserve.

What Really Happened Out There in Glorietta Canyon?

There are times when dollops of discomfort morph into caldrons of terror. Or perhaps there are just ill-fated journeys, though my optimistic and scientific daytime mind tends to reject that hypothesis. It is only in the dark, wee hours, when thoughts of mortality can come riding in like they were invited, that I wonder about how one's life might be fated. Certainly, my strangest, mind-bogglingly bizarre, and most terrifying camping misadventure occurred in 1997 in Southern California's Anza Borrego State Park. If I subscribed to the ill-fated-trip mentality, there would be evidence to support it. My day planner entry read: "Bad start to strange trip." On the morning of the day that Dave and I were due to fly out West to explore the Mohave Desert in springtime, some likely juvenile delinquents slashed my state truck tires, along with tires of other vehicles on my street in Gainesville, Florida. I had to wait around for the police and then coordinate with one of my colleagues to have the truck towed and the tires replaced while I was on vacation.

Despite the somewhat ominous prelude to the trip, I was stoked to be heading to Anza Borrego, California's largest state park, named for both Spanish explorer Juan Bautista de Anza and for the Spanish word *borrego*, or bighorn sheep. I could hardly wait to see the badlands and bajadas (alluvial fans at the base of mountains), the towering ocotillo and agave, and the sheer vastness of this highly touted Mohave Desert locale, so large that it has some five hundred miles of dirt roads.

Because April 1997 was the time that the Hale-Bopp comet would be visible, we also were anticipating viewing this astronomical phenomenon in the clear and open desert sky. This comet had the distinction of being the farthest comet from the sun to have been discovered by amateurs, interestingly in New Mexico and Arizona. We stopped at the visitor center and chatted with a ranger about best places to camp and view the comet. The ranger suggested we try Glorietta Canyon, south of the town of Borrego Springs, down a two-mile, narrow, and remote dirt road. I reveled in the vistas as we made our way to the end of the road, parked, and hiked over a hill to set up camp. We ate dinner and were enjoying the twilight, always my favorite time of day, when we heard vehicles coming down the dirt road.

Though I didn't know it at the time, we would learn a valuable lesson: Never camp at the end of a road, especially in remote locales. At first, I was just bummed and hoped that these folks, whoever they might be, were just touring the area and would drive back out. We couldn't see them, because our camp was on the far side of a small hill from where we'd left the car. But much to my dismay, because I highly value privacy and solace when I camp, they seemed to be settling in. It was, after all, a gorgeous Saturday night in a renowned state park.

Dave tends to be much more gregarious than I am and, at first, wasn't necessarily concerned that other campers might be nearby. But as the minutes ticked by and darkness began to descend, it became quite obvious that these were not ordinary campers. They were exceedingly loud and raucous. Great, I thought, we're camping near inconsiderate jerks. So initially, this scenario only seemed like a dollop, albeit a large dollop, of disappointment.

That assessment changed drastically when a bullet whizzed over our heads. "What the hell!" I exclaimed as we dove off the prominent rock where we had been sitting. Another bullet whizzed overhead, and now we were both worried. No sane camper, upon seeing a parked vehicle and deducing that someone was likely out there, would fire a gun in that general direction. The sound of breaking bottles told us that these idiots were definitely getting drunk. We, of course, had no idea that the situation would deteriorate so precipitously, and we weren't about to go back to our rental car (which we hoped was still intact) and confront them. We didn't feel safe retreating to our tent, so we grabbed a water bottle and a flashlight and headed out farther away from the terminus of the road. In retrospect, I guess we thought that maybe they would finish partying and eventually things would settle down enough for us to go back to our tent and get some sleep. But their perplexing actions and behavior only became more bizarre. Firearms, fireworks, and chanting all came into play. And the chanting was not the rhythmic Native American version: It was downright eerie and worrisome, almost demonic in nature.

As is typical in the desert after the sun sets, it was getting cold, and we considered our options. We could circle around and try to hike back toward town, but it would be an ankle-twisting struggle over rugged, rocky, cacti-strewn mountains and through arroyos (washes). This was a decidedly dissected landscape. I had recalled seeing some RVs on the road in, but by now it was getting late, and that option would require detouring around the cult (or whatever they were) and probably scaring folks sleeping in their campers. There seemed to be a break in the firearm action, so Dave left me hiding behind some boulders and quickly hiked back toward the tent to get the sleeping bags. A coyote yipped nearby, and that sound alerted him that these very strange people were also approaching our tent from the other direction. He froze and heard them say "Here's their tent; get up here." Truly, no one in his or her right mind would approach a tent that way in the darkness.

Dave waited, and when he heard them retreat, he grabbed the bags and more water and came back to me. There was nothing of real value in the tent and nothing seemed to be missing, so if robbery was their motive, they were thwarted. When he told me what he had heard, I now realized that we were well beyond mere inconvenience and were experiencing something much more sinister. This situation was akin to a Hollywood B-movie with horror overtones, and I wondered if we would be killed and become tomorrow's news. I later wrote: "longest, most terrifying night of my life." We did see the Hale-Bopp comet as a flash in the night sky, but even that sighting was overshadowed by this bizarre situation that had been thrust upon us.

All through the night and into the wee hours, we hunkered down in our bags in the sandy spot behind the rocks, as these people wandered around within earshot. What were they doing, and what were they looking for? Would they harm or kill us? I tried to take some solace when I heard a woman's voice among the male noise, but then again, some women are sucked into cults, witchcraft, or satanic rituals. Needless to say, I didn't sleep but laid there and prayed they wouldn't find us.

At dawn, I really had to pee, which of course required me to get up and move around. Dave didn't think that was a good idea, but by now, I was emotionally exhausted and, well, angry. I stood up and saw two dark figures sitting up on one of the nearby hills, watching. At that point, I told Dave that I didn't care if they saw me pee. Things were quiet, and no one seemed to be up and about except the two watchers. We decided we had to get out of there, so we went back and packed up our tent, and cautiously made our way toward our car.

People were sleeping all over the ground, and the campsite was a complete mess. We threaded our way through the sound sleepers in their likely drunken stupors, threw our gear in the car (which thankfully appeared to be unassaulted), and pulled out as quietly as possible. As soon as we got off the long dirt road, we headed into town and made a report at the sheriff's office. Minimally staffed on a Sunday morning, they promised they would coordinate with the park and look into the situation when they could get adequate backup.

After I got home and decompressed, I made a call to Anza Borrego, told the person I was a wildlife biologist working for a state agency in Florida, and explained in great detail what we had endured in the park. I felt that I got the politically correct response: They didn't find firearms (well, that is probably what the watchers were doing on that hill—hiding the guns because they had to know we would report them), and these were really stupid urbanites who had totally trashed the campsite. Of course, there was no way to know, and I will never know, if these were just drunken morons or whether their behavior indicated something much more insidious. They definitely acted in an atypical, bizarre, and potentially threatening manner. I will say that my colleagues, most of whom were avid campers and hikers, were shocked when they heard the story and were thankful that Dave and I had made it through that long and dreadful night.

In a totally unanticipated postscript to this strange 1997 event, I learned of the rare 2017 wildflower bloom bonanza in the Southern California desert after frog-strangling winter rains set the stage for stunning spring beauty. And so off I went on a solo quick trip to Anza Borrego State Park from my home in New Mexico. I had enough sense to set up a tour with a reliable expedition outfit, and at our first stop, I had a sense of déjà vu. "Where are we?" I inquired of the guide. There was a pause while he wondered if he should let the word out about this premier flower locale. "I know, if you tell me, you'll have to kill me," I quipped.

He laughed and said, "Glorietta Canyon." For a moment, I was awestruck. The canyon on this gloriously sunlit March day bore little resemblance to the place of terror on that creepy night so long ago. The bright yellow brittlebush decorated the hillsides as if someone had orchestrated a divine rock garden. And interspersed among the dominant yellow shrubs were delicate gold poppies, orange globemallow, eye-popping magenta beavertail cactus flowers, and a host of other purple and white blooms. I told no one but the guide about my earlier experience there, and he, like my colleagues, was shocked. Apparently, he had never encountered anything even close to this mystifying incident, and we laughed that there were likely no such crazies in our nature-loving tour group. Although this wasn't quite the literal and literary denouement I would have wished for, where all loose strands are tied, questions are answered, and previous unknowns are made clear, it was nevertheless an appealing and fitting end to my Anza Borrego story.

Don't Fight Me, Girls; He's All Mine: Sailing with Captain Dave

My hubby Dave's past exemplifies his risk-taking nature and love of adrenaline-pumping adventure sports (formation skydiving and cave diving). Although I absolutely love adventure, jumping out of perfectly working planes and going into dark watery depths where escape can be challenging are just not my desired outdoor pursuits. I was certified for SCUBA in my younger days and dove on the stunning reefs of the Sea of Cortez in the 1970s. I recall the electric-blue damsel fish and a kaleidoscope of other colorful denizens that inhabited those turquoise waters that lapped on desert shores. But until I met Dave, I really hadn't done any sailing. Now, in the wisdom that hindsight often provides, I really don't think sailing would be my adventure of choice.

My maiden sailing voyage was actually on our 1991 honeymoon, and we were far from the tropical waters of the Caribbean or the Sea of Cortez; instead, we were sailing from Seattle's Puget Sound to Victoria, British Columbia, on a friend's impressive vessel. Fortunately, the sailing gods were relatively benevolent, and our only close call involved a blinding fog in the Straits of Juan de Fuca. We could hear the big freighters, but we couldn't see them, nor could they see us. Their foghorns, however, were deafening. There were tense moments until we made it safely to Vancouver Island.

As a result of generally smooth seas, favorable weather, and cavorting dolphins on my initial voyage, I was understandably lulled into a false sense of complacency and poised for another sailing adventure that year. Partners in sailing (dare I say misadventure?) Dave and his friend Eric concocted a trip to Bimini, one of the Bahamian islands. It was Eric who had taken us to Victoria on our honeymoon. Eric is savvy, sharp-witted, funny, and tenacious in his sailing adventures despite having lost a leg to bone cancer when he was younger. In this case, we would rent a sailboat in Ft. Lauderdale and depart on a late November evening.

On the voyage over, the moon was full and cast its luminescent magic on the water. The wind was light and the seas smooth. At the time, I had no way of knowing that this soothing sea was not going to last. Dave later noted that he was surprised when he awoke from a nap in the cockpit to find me at the helm. Eric had gotten sleepy, figured it was easy sailing right then, even for a neophyte, and gave me a bearing to follow on the compass. I felt a bit like Bill Murray in the comedy *What about Bob?* when he's lashed to the mast of the boat and exclaims, "I'm sailing!"

My first hint of what would be days of inclement weather occurred when we hit a squall as we were coming into Bimini. Apparently, the wind had clocked around and we were now dealing with a "norther." The result was rough seas, wind, and conditions that were colder than typical for the tropics. Unfortunately, this undesirable weather persisted throughout our trip.

But it was the voyage back to Florida that left me shaken and seasick for days afterward; moreover, the journey was truly the stuff of nightmares. Unbeknown to me, Dave and Eric were going back and forth discussing the wisdom of sailing across the Gulf Stream so soon after a norther. Dave later said he wasn't worried about himself, but he knew this would be quite stressful for me. Eric kept reassuring him that it would be fine.

We left the islands in the late afternoon, and the ordeal began several hours later. Dave and Eric weren't surprised, but I was definitely shocked to see eight- to ten-foot waves coming over the stern of the boat. By then, the wind had clocked around to the east so that we were sailing downwind with a following sea. The worst part was sailing in the dark in high wind and waves. Truly, the sound of those waves was terrifying. The only thing worse was actually seeing the towering wave in the stern lights of the boat, just before it crashed down on top of us.

I was knocked down over and over and finally just assumed the fetal position as the waves drenched me. It's no exaggeration to say that one or more of us could have been injured or killed. At one point, the boom almost knocked Dave off the boat, which would have been disastrous because it's unlikely we could have turned and retrieved him. Finally, I retreated below, which exacerbated my seasickness but kept me safer. Dave and Eric were now wearing harnesses up on deck.

As if the situation couldn't get worse, the main sail blew out, leaving us with only the smaller jib sail. Although our predicament was nowhere near that of the ill-fated freighter *Edmund Fitzgerald* on Lake Superior, I couldn't help thinking of a haunting line from troubadour Gordon Lightfoot's famous song: "Does anyone know where the love of God goes, when the waves turn the minutes to hours?" And of course, I recalled that Dave had been out in that same 1975 storm "when the gales of November came early" and took the *Edmund Fitzgerald* and its crew to a watery grave.

The minutes and hours dragged on, but I knew things were really bad when I could hear Eric trying to control the boat as it rose and sank in the deep troughs of waves—and he was softly moaning. By then, I was furious with both Dave and Eric because they had knowingly led us into a totally unnecessary and potentially disastrous scenario. Of course, Dave now downplays the seriousness of our plight, but he agrees that it wasn't the best decision to sail home so soon after a norther. Apparently, Eric asked Dave if he thought he would still be married after we reached Florida because he knew how angry I was. And if that had been our honeymoon, the marriage might well have been in jeopardy.

As I lay on a bench in the galley down below, extremely stressed and queasy, I sat up in terror as I heard Eric on the radio: "Mayday, Mayday, Mayday, we are a disabled sailboat." As we were coming into port in a squall, we saw a cruise ship that appeared to be heading directly into our haphazard path. Because Eric was having trouble maneuvering the boat, we were fearful of being swept into the big boat's path, and Eric was trying to alert the captain of that boat and any others trying to escape the inclement weather that we might not be able to avoid them.

He got no response. But thankfully, as we got closer, we determined that the cruise ship was actually anchored because of the horrid conditions, and we were able to get past it and into Ft. Lauderdale's marina in the wee hours. Dave later wryly noted that we made the Gulf Stream crossing in record time and reminded me that we were, after all, in the Bermuda Triangle (Devil's Triangle), where boats sometimes just disappear.

Despite my love of outdoor pursuits, I haven't sailed since that ill-fated trip. I guess I am more of a terrestrial adventurer and desert rat, though I am still quite fond of creeks and rivers. There are just too many other ways to have fun in the great outdoors than fighting large oceanic waves.

In our early vacations and adventures, Dave had teasingly given me the Indian name "Black Cloud," mostly related to rain events in the desert that might have compromised his golf. Perhaps there's some truth to the Black Cloud hypothesis because Dave's other Caribbean sojourns weren't nearly as weather plagued and definitely not as terrifying. Or maybe it was our synergistic propensity for misadventure. More likely, this trip's close call stemmed in part from bad decision-making by experienced sailors. One thing was certain: We would need to have complete transparency on future vacations. But many years later, it wasn't vacations that we were discussing; it was potentially changing our home base and therefore our life.

13

My Return to the Land of Enchantment

The Quest for a New Burrow

Throughout my career in Florida, and despite my dedication to conserving gopher tortoises and their habitat, an underlying and strong desire to return to New Mexico continued to pull at me. I had spent many vacations exploring the Southwest, and as the time for retirement grew nearer, I began to look more seriously at how I might emigrate like one of those female tortoises I had studied. I told friends and colleagues that perhaps I was somehow channeling the famous artist Georgia O'Keeffe. I certainly didn't have her talent, but I embodied her spirit. Upon arriving in New Mexico in the early part of the twentieth century, she had apparently exclaimed, "Well! Well! Well! This is wonderful. No one told me it was like this!" She promptly fell in love with New Mexico's enchanting landscapes and began to paint them. My love affair with New Mexico had ignited more slowly, but it too became intense as the years passed.

And so I began another bold quest: to find a new burrow in a land far from my current burrow in Florida. I had no doubt that both the search and transition would be a leap of faith. Granted, I wouldn't have to travel by horseback or in a covered wagon for many moons to some unseen locale in the Wild West. But fate and luck would need to intertwine, and indeed they did. Thankfully, Dave was on board with the search. As he tells it, I had informed him when we first met that I would be emigrating someday from Florida to the Southwest, probably New Mexico. Of course, he had thought that time was far in the distant future, but the time was nigh.

For several years, we traveled around Arizona and New Mexico during our vacation breaks, looking for a place to settle. Not unexpectedly, some scenic locales like Sedona and Santa Fe just didn't jibe with a biologist's salary and pension. And then my ex, Mike, asked why we weren't looking in a specific golf community on the eastern side of the Sandia Mountains near Albuquerque. "Well, why aren't we?" Dave inquired. I had just figured that home prices there would be out of our reach.

But the idea fanned a new flame, and in October 2012, I came out to New Mexico, engaged two friends as realtors, and began to search in earnest. Amazingly, the very first house of those that I viewed turned out to be *the* one. I just didn't realize it initially because I had little frame of reference for homes on the so-called green side of the imposing Sandias. In some ways, I was akin to a relocated tortoise that isn't initially sure a burrow is a perfect fit. Conceived by an artist, this modest abode was bright, open, and eclectic. I of course looked at several other houses, but upon my second viewing of this unexpected gem, I was hooked. So on October 25, 2012, my sixty-second birthday, I put an offer on the mountain home.

In retrospect, there was an almost mystical aspect to this search for a new burrow. The night before my birthday, I had undertaken a moonlit horseback ride in the rugged and colorful badlands southwest of Albuquerque. The waning sun shone golden through the wind-whipped dust as the three wranglers and I prepared for the ride and transported the horses to the trailhead. Darkness had descended, and the unnerving wind seemed eerily alive as we set out along the narrow ridgetops with their terrifyingly steep drop-offs. The moonlight softened this carved and undulating landscape, and I recall that the gypsum beds in the rock seemed to glow an ethereal white. To add to the otherworldly beauty and innate drama of this bucket-list adventure, the head wrangler saw something large, probably a cougar, leap over nearby boulders. One of the ranch dogs, a plucky Aussie shepherd named Cowboy, went in pursuit. Tangling with a mountain lion is never a good option. Long, tense moments passed as two of the wranglers went after the dog, and I was told to steady my horse because the head wrangler would indeed shoot any animal that posed a threat to his dog. Fortunately, the panther or shape-shifter had vanished; Cowboy and the cowboys returned with no shots fired. We made our way back through a deep, mysterious canyon where the moonlight could brush only the highest reaches of the cliffs. I marveled at how it truly seemed like we were in an old western, where Indians would suddenly appear on the ridge above us. Looking back, I have to wonder if this exciting adventure in a visually stunning landscape helped provide the shot of courage I needed to boldly forge ahead with my dream of returning to New Mexico.

Alas, as with most quests, there were still many hurdles and obstacles. No, I didn't have to slay dragons, but I had to get Dave out to New Mexico to view the house, overcome his strong concerns about its diamond-in-the-rough status (it did need a new coat of stucco), and actually go through the nearly always demanding house-buying journey. In yet another turn of fate, our community's golf course in Florida closed, leaving a golfing void for my husband, a void that helped seal the deal out West. The pieces of the puzzle finally fell into place, and early in 2013, the mountain home in faraway New Mexico was ours.

The next challenge was when to retire. Technically, I could stay until November 2015, having committed to a specific retirement option with my agency. But trying to juggle two homes wouldn't work for us very long. We flew out to New Mexico three times in 2013 and once in early 2014 to work on the house. I finally decided I would retire at the end of May 2014. Of course, there were many loose ends to tie up, both personally and professionally. Just going through three decades of paperwork, journals, books, and files in my office was a huge undertaking. Besides leaving my enigmatic tortoises, I would be leaving a group of dedicated colleagues who had become like family. At the time, I didn't realize how much I would miss them; such transitions are indeed bittersweet.

The Way West

But it was time to move on, and after going through the hectic task of packing for many weeks, we pulled out of our Florida home's driveway on a steamy morning in mid-June 2014. To add to the stress, a realtor was showing our house as we were bidding it farewell. Most of our belongings were already on their way west in a POD, but because we couldn't fit everything in that large rectangular container, Dave would be pulling a small U-Haul behind his van. Canine kid Kotee and I would follow in my SUV.

We had embarked on what would be a grueling three-day journey. The U-Haul exacerbated an already challenging undertaking. Between the wind and the big trucks, Dave was nearly blown off the road numerous times. I pumped adrenaline, and each time I would gasp, Kotee would also fret and sometimes arise from her hammock-like back seat and stare at me. She really hadn't traveled much, and now in her senior years, she was being transported across the country. Dave somehow missed the first pivotal route change, where we turned west on I-10 from I-75 in Florida. I had to call him, find a rest area, and unload Kotee to give her a break until he could rejoin us. In a nutshell, we were on the road seven hours the first day, ten hours the second day, and fourteen hours on the third day: 1,680 miles. I'm typically a truck-driving mama and can manage lengthy road trips, but I had to dig deep on that final day. I thought again of the pioneers who spent weeks and months bouncing along in their covered wagons, with the same dream of migrating west. We were quite lucky overall, and I was truly thrilled to finally be resettling in the Land of Enchantment.

The Natural Lure of New Mexico

So why Nuevo Mexico? What is the enticement, the lure, the spell? Many travel writers, local authors, photographers, and artists have plowed this fertile field. And to some degree, the attraction, like any love affair, is personal, subjective, and challenging to articulate or portray. At its essence, I am drawn to both the profound natural beauty of the landscape and the fascinating blend of cultures. Wherever I go, of course, I am intrigued by the indigenous wildlife, and that is certainly true here in New Mexico. Interestingly, some uninitiated folks back East think that the Southwest in general, New Mexico in particular, is just rocks and dirt.

Our late sister-in-law, Linda, came with us to see our impending venue in 2013, and after I had taken her to many scenic locales, I asked what she thought because she was so uncharacteristically quiet. "Looks like God started construction and forgot to finish," she opined. I didn't take offense and actually thought it was a humorous observation. The land here is often bare bones: rocks in myriad shades that go beyond mere red, brown, tan, gray, and white; they include vermilion, salmon, amber, rust, cinnamon, sepia, chestnut, buff, charcoal, slate, jet, and many others. And these amazingly shaped and tilted rocks can even become watermelon-pink as the sun sets: hence the name for the Sandia Mountains that loom up above our house.

Beyond and upon those bare bones is a wide spectrum of vegetation, much of it determined by elevation as well as geography. From the emerald and golden grasslands, creosote and cacti deserts, and dark sagebrush mesas, to the piñon-juniper woodlands that dominate my yard, and on to the towering ponderosa pines and white-barked aspens, and finally, to the stunning spruce-fir forests high on the mountains, the diversity is astounding. I am totally captivated by the myriad species of eye-catching wildflowers: the creamy tower of bell-shaped yucca blooms; the bright fuchsia of hedgehog cacti; the magenta brilliance of the taller cholla cacti; the funnel-shaped and striking white sacred datura; the delicate stalks of scarlet penstemon; the blue-purple lupine; and the dazzling yellow sunflowers. The high deserts and mountain meadows are a feast for the eyes, with blooming colors as varied as those of an array of gemstones.

Not to be outdone by the earth, New Mexico's skies are mesmerizing and ever changing. As I noted earlier, I used to lie and watch cloud shadows as a child, and there is ample opportunity for such observations in this big sky country. Like the rocks and wildflowers, the clouds are of many shades and shapes, and they drift lazily or race with wind. The clouds often cozy up to the mountains, hugging tightly or creating wispy fingers that brush the peaks. Perhaps my favorite clouds are the lenticular, rare in some areas but seen here over the mountains. These smooth, lens-shaped clouds are sometimes mistaken for UFOs. Particularly awe inspiring are the sunrises and sunsets, when the clouds are afire with burnished gold, lemon, tangerine, coral, peach, rose, hot-pink, magenta, lavender, and violet. There have been times when I truly could not believe what I was seeing. And then there are the virgas: the blue-gray, pewter, silver, amethyst, and other similarly shaded shafts of precipitation that generally evaporate before reaching the ground but hang like ethereal curtains in the distance.

No wonder O'Keeffe was so taken with this incredible landscape. I am continually enchanted with the buttes and the badlands, the deep gorges and cottonwood-lined rivers, and of course the mountains that, depending on the season, might be defined by dark evergreens interspersed with spring green or fall gold aspens or iced by the snows of winter. Where else can you find the undulating, glistening, gypsum dunes of White Sands and the otherworldly, underground limestone spectacle of Carlsbad Caverns?

Seasonality

Our New Mexico abode sits at an elevation of about sixty-six hundred feet, on a tier of foothills between the San Pedro Mountains and the backside of the much higher Sandia Mountains. The seasons here reflect the mountain weather, which is highly changeable and erratic. Many folks throughout the United States evoke their local weather by saying, "Don't like the current weather? Wait a few minutes and it will change." That may be true in Florida or elsewhere, but it is particularly evident in proximity to mountains. Seasons blend, and having a jacket handy at any time of year is practical.

We moved here in the summer, which can be hot and painfully dry until the monsoons arrive with their life-giving and blessed moisture. I adore the summers here because if one leaves the lower elevations and goes up, the heat abates. There is a way to escape the more intense temperatures; moreover, it's a dry heat (ha!). And the flower displays are magnificent. Even up at ten thousand feet in the Sandias, the delicate blooms among the granite rocks surprise and delight.

Summer and fall are the best times for hiking and camping. When folks back East ask when they should visit the Southwest, I urge them to come in late summer and early autumn. By late September, the aspens are in their golden-leaved glory, and the roadside chamisa (rabbitbrush) shrubs dominate the landscape with their eye-popping yellow flowers. Later, the darker-barked cottonwoods will take center stage with their leaves ranging from lemon-yellow to amber and butterscotch. Hiking in the Sangre de Cristo Mountains north of Santa Fe is magical, with pieces of aspen gold drifting down all around. And traveling along the Rio Grande south of Taos when the cottonwoods are in full fall color is breathtaking. I recently watched a raven do an amazing aerial somersault and then swoop upward toward the mountains. For a moment, I truly wished that I could shape-shift, join its exuberant flight, and easily access those remote, enticing aspen groves. It's unfortunate that the early Spanish conquistadors couldn't appreciate nature's autumnal gold rather than search for those nonexistent lost cities of gold.

Eventually, the snows and rime ice begin to glaze the pines, junipers, spruce, fir, and now muted or leafless aspens. Winter in the high mountains is challenging, but at my elevation, it can manifest itself in Camelot-like snowfalls where it snows at night and then the sun reappears in the morning and initiates the melting process. That's not to say that blizzards don't occur; they do. The day after Christmas 2015, we had more than eighteen inches of snow at our house and it stayed a long time. And even after we've swept or shoveled the walkways, misty snow showers or graupel (a weird, tiny, white ball form of precipitation) can sweep down from the Sandia Mountains unexpectedly. The cold is sometimes palpable, and until Dave finally installed a gas insert in our fireplace, we had to burn wood each winter evening just to stay warm. Santa Fe, always extraordinary, is particularly so when snow adorns the adobe buildings and piñon pine smoke wafts through the brisk air.

I recently took up snowshoeing (not a common activity in Florida) and was entranced by the serene beauty of the spruce, fir, and ghostly aspens as I trekked in the Enchanted Forest cross-country and snowshoe area near the mountain town of Red River. More recently, Dave and I snowshoed in the vast, awe-inspiring Valles Caldera, created by a supervolcano one and a quarter million years ago. I couldn't resist sending photos to my friends back East, asking if this could be a lost snow scene from the movie *Dr. Zhivago*.

When spring does come (and it often ebbs and flows), the blossoms on the fruit trees in Albuquerque and Santa Fe are eye-catching, but the unrelenting wind can drive you crazy. It descends for days at a time and wails like a gathering of banshees. I now understand why pioneer women in their prairie sod huts or mountain cabins lost their minds because of the pervasive wind. An old cowboy tune succinctly evokes this almost sentient movement of air: "It's a wonder the wind, it don't tear off your skin." Some of our neighbors emigrate temporarily in the spring, but we typically stick it out, knowing that the winds will eventually abate as spring transitions to summer. In any season, moisture is sacred in this often-parched land, and not to be spoken ill of, even if it's inconvenient at the moment. When I hear rain on our skylight, it's a reason to smile and whisper a prayer of thanks.

Critters Aplenty

My professional wildlife career may be behind me, but my fascination and strong affinity for wild animals burn brightly and unendingly. When we first moved into our piñon pine– and juniper-dominated homestead, I wondered if such a seemingly simple woodland could support much species diversity. I need not have worried and am amazed at the species that frequent the area or are passing through. Dave and I both delight in viewing our local mammals, birds, reptiles, and even a few amphibians. In late summer last year, we also witnessed the march of the tarantulas; one climbed up and over the stucco wall of our house as he headed south in search of a mate.

Coyotes yip and howl nearby and will stroll up our walkway nearly to the front door as they skirt around our house. Sometimes it's a twosome, perhaps siblings because they look young. On one occasion, a mother and her playful, half-grown pups cavorted in our side yard. I recently observed an individual with an amazing coloration and pattern, as if it were wearing a silver saddle on its back. One night as we were coming home, a gray fox with a splendid bushy tail raced across the road and up a hill. Our coyotes and foxes certainly should have plenty of prey because desert cottontail rabbits abound here and hop all around our yard, often standing on hind legs to feed on the larger seed block.

Aptly named, mega-eared mule deer will occasionally come up the walkway as if they want to peer in the front window at the human inhabitants. Last year, a spike buck leisurely strolled up to my suspended, artsy, front bird feeder and managed to get some seed out of the ceramic flat plate without knocking it to ground. He then drank from the lower and upper birdbaths before wandering over to see what our neighbor's yard might have for him.

Rock squirrels are common (we call them all Rocky), and we have had a number of close encounters that were both stressful and somewhat humorous. When we were working on our house in 2013, before we had actually moved, our microwave started making a strange noise. Appliance repairman Dave took out the microwave and found a mummified rodent in the vent to the outside. Little did we know this wouldn't be a one-time occurrence. Shortly after we moved, we heard scrabbling above the microwave and then a pitifully plaintive squeak. Now the squeak became a loud shriek, and we saw other rock squirrels frantically running around on our flat roof. Subadult critters, like teenagers, have a propensity for getting into mischief. And one of the subadult rock squirrels had fallen down the vent and was trying to get back to his mother and siblings.

Leaving the poor stranded munchkin there was not an option, so Dave and I donned protective gloves, took down the microwave, and peered with a flashlight into the vent. A furry tail hung down; the squirrel had squeezed into a ledge in the vent. Having smaller hands (and being the wildlife person in the family), I gently reached up, grabbed the squirrel, and quickly deposited it into a large plastic barrel. Dave took it out front and released it, whereupon it joyfully rejoined its siblings. Dave plugged up the entry hole, and we thought that was the end of rock squirrels in the house.

Not so: The worst was yet to come. Early one morning, I heard the now familiar scrabbling in the wall next to our bed. Then came the ear-splitting shriek. Could a squirrel somehow have fallen down into the wall and become trapped? Dave and I tried to locate the exact location of the sound to no avail. Despite our dilemma, Dave received an appliance repair call and had to leave. Of course, the minute he left, the squirrel began to vocalize so plaintively that I couldn't stand it. I stood on a stepladder to try to pinpoint the sound and almost fell off when the unearthly shrieks reached painful decibel levels. I called Dave and implored him to come home ASAP.

Fortunately, I had correctly located the sound because Dave had to tear apart a portion of the wall in the living room of our house. There, down in one of the alcoves of the wall, was a subadult squirrel. Luckily, he or she was in reach, and once again, I donned gloves and reached down to extract the young 'un. I wondered if it was the same subadult. After the release and assuredly joyful squirrel fam reunion, Dave patched up the plaster and repainted our wall. He then patched all openings in the parapet and canales (drainage-related structures) of our flat roof, and the mischievous squirrels had to find other hidey-holes and play-roofs. Now we can enjoy watching mama squirrel, with her swollen teats, swing like an acrobat from our bird feeders as she fills her cheek pouches with seed. And while I was writing this story, a rock squirrel came to our front feeder, stood on its hind legs, and peered in the window at me. They are indeed our neighbors and share our homestead (preferably outside the house).

We have a number of other representatives of the largest animal order, Rodentia. Who doesn't love the diminutive and comical chipmunks? They race around our backyard and climb in and out of the decorative pots and birdbaths. They also dwell higher up in the mountains, along with chittering red squirrels and striking tassel-eared squirrels. This year, for the first time, we had a tassel-eared squirrel in the so-called Tree of Life (a large piñon pine) in our backyard. This species, also known as an Abert's squirrel, generally frequents the higher reaches of the mountain. Burrowing pocket gophers, a different species from those in Florida, push mounds of dirt up in our yard, and we have seen their comical, buck-toothed visages on many occasions, especially after heavy precipitation forced them to do some house cleaning. Pack rats bedevil Dave because they bring both dog feces and cholla cacti joints into our woodpile; at least he live-traps and relocates them. But he gives no such mercy to the mice, which chew wires in his work van. And until he plugged up the holes, they also got into the garage and the roof above our bedroom, waking us up at undesirable hours. We do recycle the dead mice by placing them outside in select and obvious areas for the crows and ravens.

One of my favorite indigenous mammals is the striking long-tailed weasel. When I first glimpsed this long-bodied, agile creature in my backyard, I had to do a double take: Was I seeing a rare black-footed ferret? My scientific mind said "no way," and then as I looked more closely, I saw that it was a weasel. With its chestnut back, buff belly, and black face mask and tail tip, this is one handsome critter. It seemed to be searching around our flower pot and then scooted under the fence. A few months later, on a blistering hot afternoon, I noticed several subadult weasels in the backyard; one was having trouble getting under our fence to reunite with its mother and siblings. It raced up and down, squeaking frantically, and I actually got a photo of it panting. Dave quickly dug a hole under the fence, but by then it had successfully found another exit. The physical beauty of these animals belies the fact that they are voracious predators, relentless when in pursuit of their prey.

Larger predators that occur in our montane foothills include the bobcat, cougar, and black bear. A neighbor on the next street went into her garage one day and saw a bobcat peering down at her, but I haven't been so lucky. And one of my friends was fortunate enough to see a cougar lope across a road not far from our house.

Several years ago, we found long, deep scratches on the inside of one of our wooden gates and initially wondered about their origin. After dismissing wild felids and shape-shifters, we developed a more realistic hypothesis to solve the mystery of the creatively distressed gate: a bruin. In 2017, we had our first known visits by a black bear. On a warm June night, I was in bed and heard someone, or something, moving the heavy meat smoker on the back porch. Dave was still up and turned on the porch light. He saw nothing in the gloom, but he heard a loud thud and then the sound of a galloping (or galumphing) *big* creature. There was minimal damage to a bird feeder, but our fence was somewhat squashed as the frightened bear went over it and left its telltale tufts of hair. Although it's good to put the fear of God into these marauding bruins, we learned that a nonspooked bear makes less of a mess of the fence. He or she simply climbs up and over in a leisurely manner. On the second occasion in the wee hours of a July night, we slept through the visit but found multiple bird feeders damaged or destroyed, and a relatively large branch on the Tree of Life had been broken. Finally, we officially documented a bear last year via our new trail cam; alas, Dave set the camera too low, so we captured only its ample posterior and sturdy legs. During late spring and summer, we now bring in all bird feeders each evening and put them out again in the morning.

Although we haven't actually yet laid eyes on our yard bears, Dave and I had a thrilling sighting up in O'Keeffe's red rock country near Abiquiu. We were coming out a rough dirt road at dusk after a hike on the Continental Divide trail, and a young black bear lumbered across the road after a likely drink or dunk in the adjacent Chama River. We were truly impressed by the ease with which it charged up an incredibly steep slope.

Other majestic and charismatic megafauna are found in New Mexico, including elk, bighorn sheep, and pronghorn. Many years ago, I had bighorn ewes come into camp when I was backpacking, and, more recently, I was fortunate to see five rams make their way down into the precipitous Rio Grande Gorge near Taos. Last spring, I photographed a bachelor herd lounging like barroom buddies along a mountain highway. Long before we moved, when Dave and I were driving through the plains in southwestern New Mexico, I hollered at him to stop because I saw a herd of pronghorn. He thought I was saying "prawn" and figured I'd lost my mind because there was no water in sight for these delectable crustaceans.

Other mammalian so-called mesofauna (because they are mid-sized) include raccoons, badgers, skunks, and the prickly porcupine. Twice now, I have seen these quilled critters snoozing up in cottonwoods in the bosque (forest) along the Rio Grande. Jackrabbits so large that Dave has mistaken them for small deer on the golf course and playful prairie dogs that swiss-cheese the grasslands are also denizens of the lower elevations in this biologically diverse state.

After living for so many years in subtropical Florida, an avian (bird) paradise, I was not necessarily expecting the bird diversity in our neighborhood. I had spent much of my time in the Southeast looking down for snakes or peering into the dark recesses of tortoise burrows. I must confess that I have enjoyed looking up into the trees more these days and was amazed to learn that New Mexico ranks quite high overall in avian diversity. We of course have the bird version of food courts set up in various locations around our house and maintain at least three birdbaths. I never tire of watching birds bathe. There is something both endearing and humorous in their happy-flapping bathing antics. Many species reside or pass through this woodland habitat. More common species include house finches, robins, white-winged doves, crows, ravens, western bluebirds, canyon and spotted towhees, and the pervasive (in winter) dark-eyed juncos (or *munchos* as Dave calls them because they partake of our seed and suet with impressive frequency). The highest point of the mountains above our home, Sandia Crest, becomes a magnet for serious birders during the colder months because three species of rare rosy finches overwinter here.

The birdbaths are a hit with nearly all the species, but the robins and bluebirds don't come to the feeders, so we see them in the baths or on the fence. One winter, the "water-holic" robins came in such staggering numbers during February and early March that I was refilling the birdbaths five or six times a day. The bluebirds also nest in the uptown house we provided for them. Pine siskins swing from our thistle feeders, which also draw colorful lesser and American goldfinches. Nuthatches (white- and red-breasted), mountain chickadees, juniper titmice, chipping sparrows, and three woodpecker species (ladderback, hairy, and downy) also come to feeders on our Tree of Life. Perhaps my favorites are the tittering flocks of Lilliputian gray bushtits that seem almost fearless when I'm out filling feeders or photographing. Of course, all the small birds scatter when the impressive Cooper's hawk glides through the yard like a stealth bomber.

Scrub jays are common here, unlike in Florida where a different species is imperiled. As I write these words, I can see what was once deemed a western scrub jay, now a Woodhouse's scrub jay, scarfing down the peanuts in my winter blend bird seed. When we first moved here, we naïvely left the back door open to allow fresh air and the dog to enter. One morning when Dave was out walking the dog, I heard a frantic scrabbling and squawking in the dining room: Much to my horror, a scrub jay had entered the house and was beating itself against the windows and high ceilings. What to do? I grabbed a towel, and when it knocked itself a bit silly and flapped down on the dining table, I threw the towel over it and gently placed it in the shade outside. Thankfully, it eventually regained its senses and flew away. Large, gregarious, and noisy flocks of piñon jays descend on our feeders, and last winter we observed the stylishly crested Steller's jays that generally reside higher up on the mountain. But I was truly shocked one day to see a blue jay on our fence. Common in Florida, they aren't seen in these parts much.

Other favorites are seasonal and include the stunning cedar waxwings; black-headed, blue, and evening grosbeaks; Bullock's orioles; and the equally impressive western tanagers. I learned a wildlife reality lesson about those bright orange-red, yellow, and black tanagers one August morning. Dave and I had stopped at a winery on our way home from a hiking adventure among carpets of wildflowers in the Valles Caldera, and after tasting a delicious Riesling wine, I walked around to photograph this scenic locale near the town of Jemez Springs. A subadult (do you see a trend here?) western tanager was caught in some netting that encircled the grapes, and I hurried back to the winery owner to request some scissors to free this bird. She looked at me like I was loco and said, "You can't cut my nets."

As I was to learn, the reason they have the netting is to deter western tanagers and other avian fruit eaters from consuming their valuable grapes. I'm not sure what she would have done if I hadn't been there, but she (somewhat begrudgingly) followed me out to the frantic bird and we freed it. The lesson is that one man's (or woman's) admired bird species may be another man's nuisance or problem species. I certainly knew that from my work in Florida, and now I had learned about comparable wildlife issues for grape growers in New Mexico.

In summer, I do become what one of my local friends calls "a hummingbird slave." I am constantly making sugar water and filling the feeders for the pugnacious and brilliant copper rufous hummingbirds as well as the broad-tailed and black-chinned species. They squeak and dive and fight all around us and our house.

Another favorite of mine is our state bird, the roadrunner. While adult roadrunners are streamlined and agile, I had to do a double take when I saw a relatively large, plump, fluff-feathered, younger version sheltering under a juniper. Only its eye and beak looked like the real deal. And one morning as I sat at my writing desk, my eye caught a mature roadrunner chasing a lizard around a bush. For a moment, I was transported to another time and place and was watching the velociraptor chase the children in the movie *Jurassic Park*.

Similarly, the calls of our young, fledged ravens each spring are so primitive as to be reminiscent of pterodactyls, and we laughingly say that the "dactyls" are here. They noisily bounce around on our roof, leap like Irish step dancers, mischievously pull flowers from container pots, and are especially comical when begging food from their parents. Although I realize that it's anthropomorphic to say this, I swear I can see the exasperation on Mom or Dad's face as they try to help these goofy, growing kids eat the proffered mouse, peanut, or other tidbit. Other naturalists and writers, including New Mexico's own literary treasure John Nichols, have also noted the demands that beleaguered raven parents face.

These montane realms obviously do not support the diversity of herptiles that I observed in Florida and Georgia, but we do have some reptiles and even a Woodhouse's toad around our house. During the warmer months, whiptails race feverishly about, and the prairie lizards climb our stucco walls. The lucky ones that escape predation by modern-day velociraptors are often missing tails temporarily. The bright blue tails of the young Great Plain skinks are definitely enticing to would-be predators. If I had to name a favorite lizard, it would be the short-horned lizard, also called the horny toad. There is something endearing about these cryptic, blunt-faced, ant-eating little guys (and gals).

When we first moved, a young bull snake kept trying to enter our garage. Prairie and western diamondback rattlesnakes do occur in our community but are less commonly seen except when DOR. However, they do pose a potential threat to dogs and horses in the surrounding ranchettes; our mobile vet stays busy dealing with snakebites during the summer.

In the scarce wetlands, giant bullfrogs sit like princes awaiting their transformative kiss. Freshwater turtles of various species are found in the Rio Grande and other waterways. I always enjoy watching them bask with outstretched legs at our nature center in the bosque.

Alas, no desert tortoises occur here, but I can go over to southern Arizona if I need a tortoise fix. During an earlier camping trip there, I still recall being amazed at finding a desert tortoise on a steep, rocky slope where I had to use my hands to keep from falling. And the rare Bolson tortoise from Mexico is being bred and repatriated here in southern New Mexico on Ted Turner's vast ranch holdings.

"Narrow, Steep Road Ahead"

Not surprisingly, I've engaged in many outdoor explorations and adventures since moving back to New Mexico. One adventure stands out as a learning experience, proving that even retired field biologists can be taught. In July 2016, my colorful undergrad college roommate, Gail, cryptically texted me that she and her hubby, Mark (both retired docs), would be traveling through New Mexico with like-minded adventurers. Could I meet her somewhere along their route? Because cell phone coverage was spotty or nonexistent, the texts left many questions unanswered, and I didn't learn until later how this group was traveling or that they would be attempting to follow the Continental Divide from Mexico to Canada. Once Gail and I finally established a rendezvous location via a land line call from a country grocery mart, I excitedly began to pull out camping gear. Dave decided he would pass on this one. That was fortuitous because, as I later told Gail, we might have gotten divorced over the abuse I unintentionally inflicted upon my SUV.

On a hot, mid-July morning, I set out for the San Mateo Mountains, southwest of Albuquerque and basically in the middle of nowhere. The closest village is Magdalena, and the closest town is Socorro. I stopped briefly to photograph the VLA, Very Large

Array, a series of huge radio telescopes made famous by the movie *Contact*, with Jodie Foster. My impending problems ensued from a previous forest fire that had caused the suitable forest road into the rendezvous point to be closed. The Forest Service employee that I queried via phone before my journey hadn't fully answered my questions about the alternative route, an absence of salient information that could have proved disastrous.

I blissfully left Highway 60 and headed down a wide, easily navigable, gravel road, stopping to photograph the sweeping, straw-colored plains of San Agustin. It had been a hot, dry spring, with several devastating wildfires around the state. The monsoons had not yet arrived, so blooms were sparse. But I was heartened to see a delicate blush-pink flower, Rocky Mountain bee plant, waving in the breeze, and stopped again to get a close-up photo of its antenna-like stamens. Even after turning down the alternative forest road, I naïvely motored along, dodging or straddling old ruts, unaware of what lay ahead. Perhaps that is a metaphor for life: we generally don't know what lies ahead, but we forge ahead anyway. I wasn't even deterred by the large rocks or pieces of wood in the road; I simply navigated around them or got out and moved them. Hey, I thought, I'm a field biologist, and this is par for the course. For nearly two hours, I saw no other vehicles, which was somewhat of a blessing considering the road conditions.

Eventually, I was up in ponderosa pine habitat, where the road undulated and rose and fell with more frequency. Shortly thereafter, I began to worry: the road was becoming both steep and narrow, and my Toyota Highlander struggled to gain purchase on the slick, rough, rocky terrain. I have driven many back roads in my life and career and am usually a big fan of the "road less traveled." But my blood chilled when I saw a sign that had the audacity to note "narrow, steep road ahead." What the . . . ? How could the road get any narrower or steeper?

By now, the drop-offs were precipitous and terrifying. If I had met another vehicle, there was no way I was moving over. And I wasn't stopping to move rocks, either; I just bumped over them and gasped when I heard them scrape the undercarriage. The term "white knuckle" was beyond apropos in this case: my hands ached from the vice grip that I had on the steering wheel. As the minutes and miles dragged by, I alternatively cussed and prayed and summoned the aid of my guardian angels. At one point, I really thought that my car was going to just slide down the hill out of control because of the seemingly ever-moving carpet of loose rocks. Eventually, the road improved ever so slightly, and I was able to turn off that truly dreadful track onto the southern part of the road that I should have been able to use all along. I arrived shaken and thankful at East Bear Trap Campground. This lovely venue was high in the mountains, with tall spruce and fir all around the open-meadow campsite.

Before I could fully catch my breath, I spied a huge military-style vehicle approaching the campground. Wow, I thought, the Forest Service surely has gotten uptown with its vehicles. I watched a bit warily as this humongous conveyance rolled right up to me, and much to my amazement, my roomie Gail slid down from the passenger's high seat. As I was to learn, this monstrous vehicle was a Unimog, a German uber-extreme-adventure, all-wheel-drive truck that was indeed originally designed for military use.

After hugging Gail and Mark and expressing my relief over not having rolled my car off that wretched excuse for a road (obviously, they had come a different way), my eyes widened as a number of other Unimogs circled around us. I was to get an education on these amazing and extreme RVs. As folks were getting settled, I set up my tiny backpacking tent off to the side and near the start of a trail. Somehow, sleeping right next to those megatrucks wasn't reassuring.

By now, the Unimog gang had gathered around in a circle of camp chairs, with their libations of choice. And what a colorful and adventure-seeking group of fellow travelers. Gail introduced me and of course embarrassed me with her glowing praise of my accomplishments as a wildlife biologist. We had been biology majors together at Murray State and have been each other's cheerleaders for many decades. In fact, when telling my friends about her career as a "blade" (surgeon), I often recount how I once assisted her on a colonoscopy.

That evening, we had a campfire (fortunately, the group had

checked to make sure it was legal), and as a former firefighter, I made sure it was doused sufficiently the next day before they broke camp. It was a gorgeous July evening, with a half-moon rising and stars shining through the towering conifers. I don't know if it was the sugary mang-o-ritas that we imbibed, the stimulating conversation, the adrenaline of the earlier drive, or a combination, but deep sleep just wasn't going to be mine that night. A more apt description would be "creative dozing." I got up to pee, apparently on a slight incline because I toppled over backward while trying to get in position and had to stifle laughs about my innate clumsiness. By then, the stars were stunning, but it had cooled off considerably, and my down sleeping bag beckoned.

Both the Unimog clan and I needed to head out the next morning. Although I had to detour far to the south, the road was much better except for one dicey section. Fortunately, the scary part was brief, and I knew that the considerably broader Unimogs had made it through without incident the previous day. As I got back on the wide gravel road (which seemed like a superhighway by comparison) and followed it northward, I was thrilled when a pronghorn darted across the road and then paused and turned for a photo op. Now this was a fitting way to end the back-roads part of this brief but memorable adventure. Still I will never drive that slice-of-hell forest road again.

The opportunities for wildlife observation and outdoor adventures are endless here in New Mexico. Each exploration only whets the appetite for more. Whether hiking, snowshoeing, riding, camping, bird-watching, or photographing, I am thrilled to have returned to my *querencia*. This Spanish metaphysical concept denotes the place where one's strength of character is drawn, where one feels at home. I know that many more adventures await, most embodied in explorations of the natural world, some stimulated by art and culture, and perhaps even a few more that embrace the literary realm.

As I reflect on both the meandering and straight-arrow trails that led me to this point in my life, I realize that the pathways taken resulted from a rich brew of serendipity, luck, timing, creativity, human relationships (both personal and professional), geographic pull, and just downright grit. The gopher tortoise and I were a good fit because we both exemplify that sense of unwavering tenacity.

In a way, my life became a tale of "South by Southwest," for although I grew up near the nation's capital, most of my life has really been shaped in the southern states or in the Land of Enchantment. Encounters with colorful characters, both human and animal, have enriched my life and served as fodder for my stories. And the love of animals, a passion for adventure, and a powerful thirst for knowledge are never-ending thematic ribbons running through my life.

Never having set out to be a pioneer, I remain heartened to know that when I retired, I passed the baton to other women working with wildlife. My hope is that this memoir will prove inspirational as they chart their unique paths to conserve and manage our precious natural resources.

Index

adventures, JB: badlands horseback ride, 211–212; cactus-fused fingers, 196–198; Camp Blanding Military Reservation, 155–156; Cessna ride, ix–x, 124; El Pinacate, 194–201; gators, 172–174; Glorietta Canyon, 201–206; Mexico border, 199–201; sailing, 206–209; San Mateo Mountains, 225–228; scallop gathering, 132–133; travel, reasons for, 193–194; wildflowers, 205–206
Alabama Cooperative Wildlife Research Unit at Auburn, 57, 64–65. *See also* eastern indigo snake distribution study
alligators, 172–174
alligator trappers, 105
Anatolian shepherds (dog breed), 185
Anza Borrego State Park, 201–206
athersclerosis research, 26–27
Auburn, Alabama, 61–62, 68
Auburn University, 59, 61, 64–65, 69

bear, 8, 115–116, 175–177, 185
Benji (dog), 62, 178
Berish, Joan Diemer: awards, 161; childhood, 2–9; desires and dreams, 2, 7; education, 17–19, 35, 55, 57, 61, 64–65; health, 54, 139, 146, 162, 182, 197–198; horses and, 10–17, 45; the law and, 16–17, 86–88; Manzanita Mountain house, 36–37, 55; nicknames, 48, 70, 110, 182, 209. *See also* career, JB; dogs, JB's; helitack firefighter, JB

Berish, Joan Diemer, personal life: childhood, 2–9, 16–17, 214; family, 3, 8–10, 12–13, 15, 24, 93, 132–133; husband and marriage, first, 9, 23–24, 29–30, 36, 48, 51, 53, 54–55, 61–62, 91–93, 92–93, 211; husband and marriage, second, 132–137, 168, 179–181, 185–189, 200, 202–209, 210–213, 217–222, 225; love, looking for, 131–134; retirement, 212
Bermuda Triangle, 209
bighorn sheep, 221
bigotry, 49–50
Big Shoals Wildlife Management Area, 143, 147
biodiversity, 165
black racer, 60, 76
blue-footed booby, 207*p*
Bobo (buzzard), 68–69
Bolson tortoise, 163
bosses: impossible, 28–30, 53–54; unusual, 64–68; walk-talks with, 38

cactus, 196–198, 200–201
Camp Blanding Military Reservation, 155–156
canebrakes, 168
Caras, Roger, 178
career. *See also* helitack firefighter, JB
career, JB: Alabama Cooperative Wildlife Research Unit at Auburn, 59, 64–65; beginnings, 19–20, 25; bosses, 31–35; jobs vs., 25; marriage and, 24; nuclear medicine technician, 26–28; preparing for

a, 34, 36; sexual harassment, 28–30, 84–86; University of New Mexico School of Medicine at the Veterans Administration (VA) Hospital, 26–30; USPS, working at the, 20–23; veterinary technician, 31–35. *See also* eastern indigo snake distribution study; gopher tortoise research study; helitack firefighter
cat bites, 33
cats, aggressive, 33
Cecil Field Naval Air Station, 143–146, 149–150
Cessna airplane, ix–x, 124
cholla cactus, 198, 200
Comanche (goldfish), 4
Comanche (horse), 10
Continental Divide, 221, 225, 228
Coolidge, Calvin, 57
Corwin, Jeff, 176
cosmetic surgery on dogs, 34
Cowboy (dog), 211
coydogs, 177
coyote, 197–198, 217
crackers, 78–79, 106
Craighead, John, 75
Crews, Harry, 74

Dalton, Lacy J., 183
diamondback rattlesnake, 66, 73, 168, 169, 181, 200
Diemer, Joan E.. *See* Berish, Joan Diemer
dog bites, 22–23, 33
dogs: aggressive, 32–33; conservation K-9s, 184–186; dew claws, removing, 33–34; ear crops, 34; escape artists, 188; euthanizing, 33; injured, 187–188; leg amputations, 32; tail bobbing, 33–34; tortoises and, 186–187
dogs, JB's: Benji, 62, 178; childhood, 3; Elsa, 62, 178; Kotee (Lakota), 179–183, 213; love of, 177–178; Merlin, 179–183; Randy (dog), 3, 178; Stormin' Norman, 184–189; travelling with, 62; Woofy, 3
Don Juan (gopher), 137
Dufus (tortoise), ix–x, 121–125, 138

Edmund Fitzgerald, 134, 208
Eglin Air Force Base, 103, 142–143, 164
Egmont Key tortoises, 142
Elsa (dog), 62, 178
ethnozoology, 61

A Feast of Snakes (Crew), 74
Flicka (horse), 12–17
Florida Game and Fresh Water Fish Commission, 94–98
forestry practices, 129–130
Fowler, Jim, 176
fox, 4, 217
Foxworthy, Jeff, 78, 106

Gainesville Wildlife Research Lab, 106. *See also* gopher tortoise research
Galápagos tortoise, 163
Garden Gate Nursery, 13–17
geometric tortoises, 184
Glorietta Canyon, 201–206
Gold Head Branch State Park, 143–145, 150
Goodall, Jane, 5, 23
gopher frogs, 155–156
gopher pullers, 98–106, 162
gopher tortoise burrows: animals inhabiting, 165–166; bears and, 115; cohabitation in, 121; described, 89, 108, 111–112, 120–121; digging out post-collapse, 129–130; excavating, 159–160, 162; gasoline in, 60, 73, 88; importance of, 95–96; locating, 89; rain and, 113; shape of, 97;

snakes in, 60, 70, 72–74, 77, 80–81, 88–89, 94
Gopher Tortoise Council (GTC), 98
gopher tortoise eggs, 128–129, 136
gopher tortoise races, 153–154
gopher tortoise research study: adventures, 108–110, 114–115; blood, drawing, 139–143; colleagues, 106–111, 160–163; Dufus (tortoise), 121–125, 138; entomological annoyances, 117–118; information, gathering, 97; management plan, 163–166; nostril flushing, 143–144; relocations, 150–160; suffering for science, 101, 109–110, 113, 116, 116–118, 142; traps, setting, 109, 111–113; weather and, 138–139, 148–149
gopher tortoise research study areas: Big Shoals Wildlife Management Area, 143, 147; Camp Blanding Military Reservation, 155–156; Cecil Field Naval Air Station, 143–146, 149–150; follow-ups, 135–138; Gold Head Branch State Park, 143–145, 150; Lochloosa Wildlife Management Area, 119–121, 128, 135–138; Oldenburg Mitigation Park, 143–144, 146–147, 150; Roberts Ranch, 113–118, 136; sandhill study site, 113–118; Wolfe's Pasture, 123–124, 138
gopher tortoises: anthropogenic impacts, 96, 165; behavior plasticity, 126; clutch size, 115; described, 115, 123; dogs and, 186–187; emigration, 121–127; females, 125–129; habitat, 66, 96, 121–127; harvesting, 98–106, 142; managing, 163–166; mortality events, 139–140; penises, 121–122; recaptures, 137–138; relocating, 150–160; restraining, 141; selling, 102; size of, 115, 123; threats to, 96, 97, 103–104, 129, 139–150, 151–153, 164–165; tracking, 119–125; trapping, 111–113, 136, 138; URTD, 139–150; waifs, 164
gopher tortoise stew, 102
gopher tortoise ticks, 118
Grand Canyon, 49
Gretchen (dog), 94
Grogan, John, 185

#MeToo movement, 29
heldt, 70
helitack firefighter, JB: in Arizona, 48–52; bosses, 38; commute to and from Grants, 36–37, 47; crew members, 37–39; first fire, 43–44; health, 54; recreation, 51–52
helitack firefighters: helicopters, 39; meals, 41, 43, 44, 45–46, 49, 50, 54; physical conditioning, 40–42, 54; training, 39–43
helitack firefighting: mopping up, 42–43, 47–48, 54; slurry drops, 40, 45–46, 49–51
Hill, Anita, 29
Hurricane Georges, 147

indigo plant, 104
indigo snake bites, 107
indigo snake distribution study: colleagues, 66–67, 70; findings, 90–91; funding, 59; habitat, 74–75; inaugural study, 71–74; information, procuring local, 72–73, 78–83, 86, 88–89; law officers, dealing with, 84–85; Naval Submarine Support Base search, 89–90; palmetto flatwoods, 88–90; presentations, 171–172; snake ladies, 83–84; suffering for science, 71, 82, 89–90; technicians, 69–71
indigo snakes: about, 60–61; common names, 60; dead-on-road, 76–77; in gopher tortoise burrows, 60, 91;

habitat, 60, 74–75; in jail, 87–88; searching for, 76–77; threats to, 60, 60–61; tracks, 77

JB, transitions: Alabama, 61–62; college, 17; Florida Game and Fresh Water Fish Commission, 93–94; to New Mexico, 24, 210–213
jokes, practical, 66–67

Kangals (dog breed), 185
Karelian bear dogs, 185
Keller, Helen, 191
Kotee (Lakota) (dog), 179–183, 213

Lake Okeechobee, 170–171
Lakota (Kotee) (dog), 179–183, 213
Landsat imagery, 74–75
Lightfoot, Gordon, 208
Little Vic (goldfish), 134
Loachapoka, Alabama, 69
Lobo Canyon fire, 45
Lochloosa Wildlife Management Area, 119–121, 128, 135–139
Lonesome George (tortoise), 163
Lost Fire, 47–48
Ludowice, Georgia, 64

MacDougal Crater, 200
Magdalena, NM, 52–54
Manzanita Mountains, 36–37
Marker, Laurie, 185
Merlin (dog), 179–183
Mexico border, 199–201
Minorcans, 104, 142
Mitzi (cat), 3–4
Molina Crater, 198
monkeys, 27
Moon Crater, 197
Mount Elden, AZ, 50
Mount Taylor, 41–42, 44, 46–47, 52
mule deer, 218

Naval Submarine Support Base Kings Bay, 89–90
Nelson, Willie, 31–32
New Mexico: animal diversity, 217–225; birds, 222–224; JB transitions to, 24, 210–225, 228–229; landscape, 213–216; lure of, 213–215; seasonality, 215–217; skies and clouds, 214; weather, 215; the wind, 217
Nichols, John, 224

O'Keefe, Georgia, 64, 210, 215
Okefenokee Joe, 78–79
Oldenburg Mitigation Park, 143–144, 146–147, 150
Opelika, Alabama, 62–63

Papago Tanks, 200
Parker, Megan, 184
Perkins, Marlin, 176
Phillips Buttes, 201
El Pinacate, 194–201
Princess Norkett (horse), 10

querencia, 228

rabbits, 4, 26, 217, 221
radiation safety, 27–28
Radio fire, 50–51
rain, 21, 138–139; burrow collapses, 130; camping and, 196; capturing tortoises in the, 112, 138–139, 149; desert, 196, 209, 217; dogs and, 179; firefighting, 46, 51; serious, 114, 117, 148–149; virga, 215
rain catchers, 200
Randy (dog), 3, 178
rats, 26, 219
rattlesnake roundups, 60, 73–74, 80–82, 84, 86–87
rattlesnakes: collecting, 167–168; hunting, 78–83; tracks, 77; venom

extraction, 73, 86. *See also specific species*
ravens, 201, 222, 224
Rawlings, Marjorie Kinnan, 102, 119
roadrunners, 224
Roberts Ranch, 113–118, 136
rock squirrels, 218–219
rodents, 219–220
Rodney (snake), 81

saguaro, 200–201
salamander, 106
Sandia Mountains, 36, 214, 215–216
Sandy (dog), 185
Sangre de Cristo Mountains, 216
San Mateo Mountains, 225–228
Santa Fe, NM, 211, 216–217
scrub jay, 222–223
sexual discrimination, 55
sexual harassment, 28–30, 84–86
Sheard, Virna, 5
Skin Shop, 167–168
snake bites, venomous, 73
snake hunters, 78–83, 86–87
snake ladies, 83–84
snakes: dead-on-road, 76–77; fear of, 59–60; in gopher tortoise burrows, 60, 70, 72–74, 77, 80–81, 88–89, 94; killing, 53, 60–61; in toilets, 69; transporting, 80, 82. *See also specific species*
snakeskin processing, 167–168
softshell turtles, 170–171
Sonoran desert tortoise, 94, 127
Sonoyta, Mexico, 199
Southeastern Wildlife Conference, 66–67
spring lizard, 106
squirrels, 187, 218–219
Stormin' Norman (dog), 184–189

Swampwise (video), 79
Sykes Crater, 201

Tapia Fire, 47
tarantulas, 217
Thomas, Clarence, 29
Thumper (rabbit), 4
ticks, 5, 118, 136
Tiny (horse), 16
Tohono O'odham (Papago), 200–201
tortoise restraint device (TRD), 141
TR (tortoise), 152
Tucker (dog), 184

Unimog, 227–228
University of New Mexico School of Medicine at the Veterans Administration (VA) Hospital, 26–30
upper respiratory tract disease (URTD), 139–150
USPS, working at the, 20–23

Valles Caldera, NM, 216, 223
veterinary medicine, 31–35
Vietnam War veterans, 28

Wayne, John, 11, 19
weasel, long-tailed, 220
wildlife biologists, truth about, 61
wildlife management prescribed fire, 164
Wolfe's Pasture, 124–125, 138
woodpecker, pileated, 106
Woofy (dog), 3

Young, Colonel, 10

Zona de Silencio, Mexico, 163
Zuni Mountains, 43, 54